Praise for:

BACKWARDS and BACKWARDS Guidebook

"*BACKWARDS* is deftly written and researched, and highly recommended."—**Midwest Book Review**

"A strong choice for those looking for answers about life, the universe, and everything, *BACKWARDS Guidebook* is a definite purchase."—**Midwest Book Review**

"Consistent with works such as Conversations with God and a Course in Miracles, BACKWARDS clarifies and simplifies the complexities that we tend to manifest into our lives."—**TCM Reviews**

"Nanci went farther than most NDErs. … She looked at life, and death, from multidimensional perspectives—the human, the soul, and Light Being, and Source. … If you read just one new book about NDEs this year, make it this one."—**Vital Signs Magazine**

"Danison's elaborate but very readable story is summarized in her first book, *BACKWARDS: Returning to Our Source for Answers.* … It's a pretty tough observation, which to me lends credibility: Danison is just telling us what she learned—no sugar-coating."—**The Spokane Spokesman-Review**

BACKWARDS Beliefs

"What a fascinating and disturbing book (in a good way)."—**Editor, Fate Magazine**

"The author sets out to compare the spiritual truths she learned during the experience with ideas from organized religions. … Readers interested in near-death experiences may find the book compelling for its depiction of reality."—**Kirkus Book Reviews**

Answers From The Afterlife

"Nanci Danison's writing is crystal clear. I have zero doubt she experienced everything she wrote about in this book."—**Valerie Gangas on Amazon.com**

"An excellent book with every question answered so eloquently. Nanci's explanations and descriptions of the afterlife, in all her books, are the best I've come across in this subject. I thank God (Source) it happened to a lawyer, she can, and does, explain the unexplainable."—**Hideko Imanishi on Amazon.com**

"Incredible! Nanci L. Danison has once again bestowed upon humanity a spiritual masterpiece. Her latest book, *Answers From The Afterlife,* expounds upon much information that can be found in her three previous books. Anyone truly interested in near-death experiences, the afterlife, and creation itself, you need not look any further.—**Alex Turner on Amazon.com**

Create a New Reality

"She simply lays out what manifesting is and is not, and how to go about manifesting in one's own life. It is a breath of fresh air to see a book on this topic that is not focused on manifesting money."—**Kevin on Amazon.com**

"So this book had been life changing for me, because the information is directly from the Source of us all."—**Casey Anderson on Amazon.com**

"I feel blessed to have come cross Nanci L. Danison's book for the amount of manifesting details I have intuitively and logically picked up on."—**Stella Carrier on Amazon.com**

33 SOULS
WHO MET GOD

Also by Nanci L. Danison

Books

BACKWARDS: Returning to Our Source for Answers
BACKWARDS Guidebook
BACKWARDS Beliefs: Revealing Eternal Truths Hidden in Religions
Answers From The Afterlife
Create a New Reality – Move Beyond Law of Attraction Theory

Videos

Our Self-Healing Superpowers
Startling Truths About Heaven from Deep NDEs
Suffering's Not a Purpose of Life
Reincarnation, Karma and Soulmates
Source's Purpose for Dreams
Angels, Guides, Ghosts, Demons and Other Spiritual Beings
Why NDEs Are All So Different
The Creation Story – What is God? What am I?
Crossing Over Process Explained
Aliens Seen in the Afterlife
What Do We Do in Heaven?
Source's Tips for Getting to Heaven
Afterlife Experiencers Confirm Past Lives
How Common Religious/Spiritual Beliefs Work
Who's to Blame for Evil
The 5 Purposes of Life
What Unconditional Love Looks Like
The Future Source Showed Me (the 3rd Epoch)
How to Access Source's Knowledge
An Afterlife Alternative to Incarnating

33 SOULS WHO MET GOD

Accounts of Atheists, Buddhists, and Christians Who Died and Met the Same God

Nanci L. Danison

A.P. Lee & Co., Ltd.
Columbus, OH

A.P. Lee & Co., Ltd.
PO Box 340292
Columbus, OH 43234-0292
"Evolving mankind, one book at a time."

Cover design: Karen McDiarmid
Cover graphic: © Chip Eggerton

Library of Congress Control Number: 2024922133
ISBN: 978-1-934482-41-4

Printed in the United States of America

Contents

Acknowledgements

SO MANY PEOPLE HAVE HELPED MAKE IT POSSIBLE for me to conduct my research and write this book that it would be impossible for me to thank each of them individually here. I pray each of you know in your heart that your love and support have kept me going and made leaving the Light bearable. Some of the many wonderful Beings of Light on Earth who have contributed to this book are: Sherrill Shaffer, Tom Hamblin, Karl Krumins, and many others who sent me NDE accounts to read as I searched for my 33 Souls Who Met God; Tom Hamblin, the articulate soul to whom I am forever indebted for reviewing my original Preface and giving me the term "expanded afterlife experiencer;" Dan Tylavsky, the retired professor who reviewed and edited my original preface and conclusion and gave me invaluable advice about the scientific aspects of the book; and Sophie Wong Lo, the logical and clear-thinking soul who reviewed and edited the last three chapters for errors, consistency, and logic. As always, I am indebted to Erin Clermont, my editor, for saving me from putting my foot into my mouth; and to my patient book and cover designer Karen McDiarmid. To Chip Eggerton, who designed the cover graphic that made me cry with longing for my home in Source, I cannot express enough gratitude to you dear heart. Special thanks are extended to the many expanded afterlife experiencers (and their publishers) who have allowed me to weave their afterlife experience accounts and words into my text. I also feel a huge amount of gratitude for the Near-Death Experience Research Foundation's online database that provided me with most of my 33 Souls.

Explosive New Afterlife Research

THE GREATEST UNCERTAINTY IN HUMAN LIFE IS: what happens after it ends? The most direct answer is available to us from those with firsthand personal knowledge. A small group of afterlife experiencers has returned to this life to reveal to us the wonders and otherworldliness of the afterlife they lived during deep, transformative excursions. The biggest secret they share is that there is nothing human about life after we leave human bodies. For example, Aaron M's astounding afterlife experience included "things that are incomparable to normal life and [he] wouldn't know how to begin describing it."* More importantly, these fonts of impeccable information obtained directly from God prove that life and love are infinitely greater, broader, and more wondrous than what we are experiencing now.

Humans naturally view life through the prism of their own shared experience and see that familiarity reflected back to them in everything they behold. Take, for instance, such commonly accepted fictions as the man in the moon, dogs smiling at us, and movie aliens that have humanoid bodies. People humanize, or anthropomorphize, something in an effort to bring it within their comfort zone of familiarity. We even have prevalent beliefs that the Supreme Deity is human in appearance and character.

Most of our religious and spiritual models of life are based on bedrock beliefs that human life is the epitome of existence and mankind the only intelligent life in the universe. Such thinking leads to the conclusion that humans are God's only children—the only creatures eligible for life after death. Even those who believe animals have souls never consider whether heaven offers animal souls the same bliss as human souls, rather than a human-centered bliss where animal souls continue to be subservient pets. Nor can animal lovers conceive of a God that looks like a dog or horse, or angels that look like cats and pigs, or a heaven filled with cars and mice for dog and cat souls to chase. Yet, they see no irony in the fact that they readily project a male human's appearance unto God, and human life qualities onto the afterlife, and consider it absolute fact. Literally no one gives thought to the idea that if aliens exist, they may have beliefs about their own afterlife. What if you had chosen to be incarnated into a non-human being or thing somewhere else in the universe? Would you think God looks like a human man with long white hair wearing white flowing robes if you had never heard of humans and had no idea what they look like? Would you believe that a humanoid God is *your* God? Would you believe that heaven is located in a realm above Earth if you had never heard of Earth and had no idea where it is? Would you want to ascend to a heaven located in such an alien place? Would you believe that heaven has earthly gardens, cities, and mansions when you have no idea what those things are? Would you feel bliss in an environment that looks totally foreign to you because it is earthlike rather than like your home planet? Or would you be afraid of it?

It is perfectly natural to project what we have learned about human life onto all aspects of life, non-human life included. After all, humans have nothing more from which to gain any perspective. But this practice ignores the vastness of the universe. Worse, painting all of life with a human persona can lead to profoundly erroneous and frightening results when it comes to life after human death. For example, because humans seek revenge against those who wrong them, our religions and spiritual belief systems may include a hell where our enemies are punished for all eternity. But what if life after death is not based on violent human emotions and traits? Is it possible that humans have projected what they have experienced among humans onto God and the afterlife although the truth about both is not that limited?

Those of us who have temporarily lived deeply within the afterlife and returned know that any belief or spiritual philosophy that applies only to human experience, and not equally to the rest of the universe, comes from man and not God. Most medical near-death experience (NDE) reports and research reinforce the notion that all of life is centered on humans. It is not.

In a sense, NDE researchers have stuck their noses into the tent of the afterlife and have reported back what NDErs observed immediately around the tent flaps, using words based on human experience. This has happened for two reasons. First, the vast majority of reported NDEs occur before crossing over into the afterlife, if at all, or at the very beginning of the afterlife. I call the souls who do briefly enter the afterlife *threshold afterlife* experiencers. These souls barely glimpse the afterlife before they are resuscitated or otherwise return to their bodies. They, of course, are unaware of what more they might have experienced had they stayed in the afterlife. More important, they continued to think like humans while in the afterlife. Very few reports detail what happens deep within the afterlife from a perspective not based on human expectations and projections. I found only 33 souls, among approximately two thousand accounts I reviewed, who not only had extensive and deep afterlife visits but they did not project human beliefs unto what they learned and experienced. Second, scientists' job is to study the material world, so they understandably limit themselves to it. Consequently, most NDE researchers have traditionally considered only those reports couched in terms of material world experiences with which humans can identify. Otherworldly or mystical experiences seem too far removed from physical life to fit within the medical realm of study.

This book focuses on reports of what exists inside the tent. What I call *expanded afterlife experiencers* have died and crossed over into the afterlife, some of them many times. The experience has become so familiar to them that they do not need the comfort of human life projections. Their reports consist of events and knowledge unknown to mankind, of life supernatural in nature. NDE researchers typically categorize these accounts as outliers because they really are outside medicine's realm of inquiry and different from the vast majority of NDE reports. Researchers have no way of knowing that these outliers are not aberrations. They reveal what lies far beyond the traditional medical model NDE. The unintended result is that science's history-making step into afterlife research has simply reinforced our misperception of the afterlife as a glorified Earth with superhuman inhabitants.

My research project takes the next step in NDE research—into the center of the afterlife tent. The profound difference between my research and that of other NDE researchers is that I have personally experienced everything I write about here. I have "been there, done that" and lived in the afterlife long past what other NDE and threshold afterlife experiencers report. Consequently, I am in the unique position, using hindsight, to review NDE accounts and determine which are those of expanded afterlife experiencers, regardless of the terms they use to describe what they saw, heard, and felt. I recognize from personal experience what other expanded afterlife experiencers are describing.

Together, we expanded afterlife experiencers offer eternal truths about life, death, the afterlife, and our Creator, untainted by the projection of human life expectations and interpretations onto those truths. Our afterlives will astonish you. You may find it too hard at first to accept God's true nature, and your own true nature and power, as we describe them. I certainly did even though I knew from firsthand, personal experience that it is the absolute truth. It took me years to accept that what the Catholic Church had taught me is simply human speculation. So, if you struggle with this text, I totally understand. I just hope our intimate stories of living in the afterlife will nevertheless give you the same life-altering comfort and peace I have finally gained from knowing the how and why consciousness survives leaving the body.

I believe the compiled knowledge in this book has the potential to not only give you a clearer understanding of what awaits you when you return to your home after this incarnation ends, but also to change your whole perspective on life. For me, living in the afterlife for up to thirty minutes once, and unknown amounts of time twice more, lifted a huge burden of anxiety and guilt from my shoulders put there by my childhood religious training. I learned during my first afterlife visit that our Creator did not shove us out into a cruel world, alone, with no way to help ourselves. We are not now, and never have been alone or isolated from the unconditional love of our Creator and each other. We are not victims of circumstances beyond our control. Nor are we held to ever-fluctuating church rules of behavior we must follow to earn our way into heaven. We are powerful beyond measure. My research into the afterlives of my fellow expanded afterlife experiencers will tell you how and why life after human death is far grander, glorious, and complex, and much richer than any human can ever conceive of simply looking around himself on this planet.

1

We Met God in Its Raw Nature

WAYNE H, A FATAL AUTO ACCIDENT VICTIM, RE-turned to human life to enlighten us about a non-physical entity he met in heaven:

> I was shown what appeared to be a flowing river. It was silver and shimmering as it flowed. The drops in the river were each a different color yet all flowed together as one body of water. Nothing gave me the impression this was actually water or a river but this is the best descriptive example that can be given of something I witnessed for which there are no words.[1]

Sue C, who left human life after a head injury, likewise met a river or sea of colors in the afterlife, which she describes as:

> At first, it was like looking at the sky, but as I became one with this sea it was more like a translucent river of stars and colors. They would breathe and undulate and seemed to be as one. They all interacted together like an ocean, with a tide, with waves—rising and falling lazily in space.[2]

An Anonymous reporter whose body died of anaphylaxis also

met something that "appeared to resemble liquid water except it looked more like clear, flowing electricity. It flowed smoothly ... and was transparent and shimmering when doing so."[3] Later in his afterlife account, Anonymous describes the entity as galaxy-sized "with plumes of blue electrical discharge being released, as if watching a thunder cloud at night, lighting up with electrical discharge."[4] Multiple death survivor Jennifer J made contact with a phenomenon she depicts as "undefined, fluid, existence, essence of life" during her afterlife visit due to multi-organ failure in her body.[5]

Another returnee from heaven, Arti Gupta (now Mira Sai), whose human life ended in an auto accident, offers a more galactic portrait of the entity: "It seemed to have a consistency of the ever-finest, minutest electric-like sparkles.... It could perhaps be compared to the light of thousands of brilliant stars reflected in millions of sparkling diamonds, all-encompassing, self-luminous, and pulsating with electric energy."[6] She adds that this energy is infinite, without beginning or end, and was both still and totally alive. Bridget F, another auto accident victim, perceived the entity as a vortex in the afterlife, a hole in the sky with clouds and lightning-like plasma around it. She returned to human life with this additional galactic imagery: "Within the hole were stars but, not the stars we see outside our atmosphere at night; rather, the center of the universe. Like a galaxy swirling around the most brilliant light one could imagine."[7] Car-truck accident victim Valeska K says it "looked like a NASA image of a galaxy ... like a white light with the spectrum of colors vibrating from it."[8]

While her body was dead ten minutes from a grand mal seizure, Malinda K ascended into heaven and encountered "an energy field, similar to the white garbled screen you would see when changing channels on a TV and finding no reception. A pulsing, constant motion of energy/electricity."[9] After my fatal bout with anaphylaxis, I made the acquaintance of "a single, unified field of energy" with a core that is "pure living, feeling energy."[10] I likened it to a sun with an expansive corona of Light and Energy encircling it and blasting outwards from it. During a Code Blue, Chantal L similarly encountered "a state of pure energy and pure love. Everything is energy on the other side," she writes.[11]

Aaron M describes experiencing a light containing thousands of organized colors in mathematical shapes and patterns that grew and glistened, pulsated, or just floated when he was in the afterlife due to a suspected drug overdose. He calls it "a force that is an endless and infinitely powerful 'thing.'"[12] Joan LH, who did not survive an anesthesia overdose, entered heaven and recognized Aaron M's "thing" as, "a force, a power–LOVE."[13] Ron Kruger, whose body died after being thrown into a windshield at 90 miles an hour, agrees with Joan's loving force-field analogy. In the afterlife, he beheld "an infinite expanse of glorious light … evenly distributed and seemed to undulate gently with a force field."[14] Head injury victim Kathryn H had awareness during her afterlife visit that "a spiritual force exists."[15] Ron further identifies this force as "the all-powerful force of Love. This love was so powerful, so extremely fulfilling–everything else was immaterial. This all-mighty power of Love goes well beyond our egotistical interpretations of the emotion."[16] DW, after a successful suicide attempt, entered a golden glow in heaven and met "an immense, golden, glowing globe shape" entity, and felt it "was composed in its very atoms of a substance I can only call love and that that substance created or was a force or power, like electricity is here."[17] She reports that what she saw next truly amazed her: "I discovered that the glowing, golden globe of light was alive. It was a 'self.' It was a living, aware, loving being."[18]

Drowning victim Demi B's expanded afterlife account reports: "I didn't see a God like being. I saw no beings but was immediately aware that there was a power there.... This energy."[19] William W, whose body also drowned, experienced "a pure love energy" in the afterlife.[20] High school party drowning victim Andy Petro uses the term "Light/God" in his account of his expanded afterlife experience. Bobby R, whose 4-year-old body stopped breathing in the emergency room, remembers, "a white light that seemed to pulsate."[21] Peter N describes a Light of absolute love, personality, intelligence and wisdom he met after leaving this world during a motorbike accident.

William Horden met "a sphere of light, but light that is aware....

As if it were One Mind,"[22] when he was in the afterlife. Yazmine S also describes what she encountered as the "One Mind," the repository of all of creation, when she entered heaven after her body bled to death.

What are all these dead people describing?

God.

That's right.

God.

Not what you expected, is it?

We souls standing in awe of our Creator struggle to apply poorly fitting earthly concepts and language to describe witnessing God's supernatural magnificence and life-altering impact on us. Famous near-death experience (NDE) researcher Dr. Bruce Greyson writes that two-thirds of his research subjects who reported meeting a divine being in heaven could not identify it. Perhaps they saw something like my research subjects saw. The God they saw was in its rawest form without the superimposed gloss of religious assumptions. It looks nothing like a human body. It has no gender. Or race. Or form. No physical attributes at all. Yet this alive Energy phenomenon ebbs, flows, undulates, or otherwise moves in some mysterious way that reminds the souls that see it of a sea, river, sun with a corona, or field of twinkling stars.

God is obviously not a literal river, sea, star field, or sun. These firsthand reporters tell us that it is not physical matter at all. It is not really a being. Nor is it accurately described dispassionately as an entity or thing. There simply are no words in any language up to the task of describing a divine phenomenon so far beyond human comprehension and experience. So other worldly. So supernatural. That is why my research subjects and I use different analogies or metaphors, like a river, sea, stars, sun, or energy source as stand-ins for accurate descriptions of this phenomenon. Yet, if you look deep enough, there is a consistency to our varied attempts to describe what we experienced. We all saw and/or felt a vast expanse of moving Energy radiating Light and love. We recognized it as the source of all love. The source of all Light. The source of all energy. And the source of everything, all of Creation. We are all trying to paint a word picture of the same thing: God, or what I refer to as the Source phenomenon or simply *Source*.

How do you know this? Because we said so.

Wayne H, a Southern Baptist, suffered his body's clinical death as the result of a head injury sustained in an automobile accident on August 24, 1996. He was pushing a disabled car in the road when he was struck from behind and pinned between the two cars. The impact was forceful enough to bend the frames of both cars. Wayne's religious background caused him to expect to be greeted in heaven by a man in white robes sitting on a golden throne, but he was not. Instead, he describes the divine this way:

> I was shown what appeared to be a flowing river. … The main body of the flow was silvery shimmering lights with different colored drops on the flow. I understood (I use this term because I did not actually hear) the colored drops were the experiences of all who had lived.… All experiences were known at once by the collective consciousness that was the stream.… That stream of consciousness and knowledge is what might be termed the "mind of God."[23]

Wayne reports he has no use for his former religion now that he has had this personal experience.

Sue C was an atheist when her body died on June 20, 2005, of a head injury. She came back into her body from the afterlife during CPR resuscitation by medical staff. She reports: "as I became one with this sea it was more like a translucent river of stars and colors.…

All I cared about was this sea of energies. I felt the entity start to explain 'This is the source.'"[24] After her visit to heaven, Sue became a Pantheist because she now knows the Source is everything.

A man who is misidentified in the Near-Death Experience Research Foundation website NDE database, resulting in my calling him "Anonymous," is an adult male with a wife and children whose human host died on July 13, 2021, from anaphylaxis due to a wasp sting. As he sat in his bedroom, Anonymous's heart rate accelerated, his nerves jangled, and he started sweating profusely. He reports feeling his blood pressure drop and his life force start to drain away. Just then, a tunnel-like formation was superimposed over the room in the vicinity of his daughter, who had come into the master bedroom to check on him. Anonymous entered the tunnel and felt a liquid water-looking clear electricity flowing from the tunnel and into the room where his body lay. He soon identified the energy:

> As I realized what I was looking at, I was absolutely floored. [I] was in the presence of the most intelligent and powerful force in this universe and beyond.
>
> This was absolute, radiating intelligence and there was no doubt about what this was.... There is no doubt this is a living entity.... Is this God I am looking at? But comparing this entity to what our earthly religion teaches us shows how infantile and naïve we humans are. Comparing this entity to God would be like comparing a 300-megaton nuclear weapon to a firecracker and saying they produce the same effects.[25]

Anonymous was a non-practicing Catholic before visiting the afterlife and says he has not affiliated with any religion since then.

Jennifer J, a Lutheran raised as a Catholic, was pronounced clinically dead multiple times while in the hospital giving birth to her fourth child. Her body had no brain activity during two additional NDEs. Jennifer bled out after delivery and during emergency surgery faster than the blood could be replaced. She had two heart attacks, a stroke, seizures, her organs shut down, and she was in a coma for weeks. She was pronounced dead, and her family was notified of her death. Jennifer sums up the nature of the deity she met: "We cannot define God, whatever we wish to call it. God isn't a person or an entity; God is undefined, fluid, existence, essence of life."[26] She has no religious beliefs after her return from the afterlife.

Arti Gupta, renamed Mira Sai, had her expanded afterlife experience on July 30, 1994, after a fatal car crash. Arti, a very successful financial consultant in San Francisco, and her secretary, were driving a Mercedes 560SL without seatbelts when their car hit a car driven by young boys who had swerved into Arti's lane and stopped. Arti accidentally hit the accelerator instead of the brake and crashed into the other car. In the afterlife, Arti saw "the ever-finest, minutest electric-like sparkles ... the light of thousands of brilliant stars reflected in millions of sparkling diamonds ... pulsating with electric energy.... I seemed to know that the light was the Supreme Infinite Light that is God, the Cosmic Consciousness."[27] Arti was Hindu before her death but was raised in a Catholic household. She also practiced Tao meditation. Now, she calls herself spiritual rather than religious. Following her experience and subsequent spiritual journey, Arti changed her name to the more widely recognized name of Mira Sai and created the Sai Foundation.

Bridget F's body died in an automobile accident in 1995 when her car rolled over several times down an embankment and landed on top of her. When she got out of body, she saw its head sticking out of the broken driver's side window. Bridget suffered seventy breaks to her left hip, fractured several left side ribs, and had ribs puncturing her left lung. Her body was bleeding out of its eyes when she returned to it and could open the eyes. Bridget had no religious background before this death but reports her beliefs to be "liberal" both before and after her expanded afterlife experience. Bridget saw a galaxy swirling around the most brilliant light and adds: "It was what I perceive to be the source.... At that point, I prostrated and said or, rather thought, oh my G-d you are!"[28] Bridget was in intensive care for thirty-six hours, where a priest administered last rites to her three times.

When Valesca K, after experiencing her body's death, saw a white light with colors vibrating in it in the afterlife she understood it was what she calls "Source."[29] She witnessed Source as the result of a complicated highway accident involving another car and then a semi-trailer truck. Valeska held an eclectic collection of spiritual beliefs both before and after her death.

Malinda K, who remembers witnessing "an energy field, similar to the white garbled screen you would see when changing channels on a TV and finding no reception" in the afterlife, calls this pulsing, constant motion of energy/electricity "a beyond vastly intelligent source of energy."[30] Malinda had suffered a grand mal seizure that stopped her body's heart and lungs for ten minutes before her husband, a police officer, revived her with CPR. Upon returning from this expanded afterlife visit, Malinda says she no longer believes in a separate omnipotent God character. The deity is energy, not a being, she says.

My first death followed an invasive radiological procedure and was caused by an anaphylactic reaction to two local anesthetic injections. While in the afterlife, I met and felt Source as an immense, infinite, living, emoting, unconditionally loving, creative, intelligent power source, or Energy field emitting a vast corona of Light that appears denser the closer I came to its core. I liken it to our sun, with a vast, deep, intense Energy core personality or essence. Source has a corona-like Energy field around it. I experienced this corona as having different colors and densities or degrees of luminosity depending upon how close to the core I came. The Energy field radiates outward, creating the less dense Energy that newly arriving souls perceive as "the Light." Source's core has definite innate character traits and personality like a being does, but it is nothing like a being. Religions refer to this core/Energy field as God, Yahweh, the Almighty, the Supreme Being, the Creator, and other names. It is pure Energy. Its Energy felt exactly like what humans call love. I was a practicing Catholic at the time of my first death. I no longer feel the need for any religious faith because now I have personal knowledge.

Chantal L's human host died during an in-hospital code blue

when a pelvic cyst the size of a grapefruit ruptured and caused septic shock. Chantal was aware that she was dying and felt herself exit the body through the crown of her head. Chantal relates that she used to be a practicing Mormon but before she died she related more to the Christian religion. She had no religion when she returned from the afterlife. Chantal writes: "This collective consciousness that I became part of, is God. Or that is the feeling I got. It is a state of pure energy and pure love. Everything is energy on the other side."[31]

Aaron M's host died on August 1, 2015, while traveling abroad with a friend. He does not know what caused death, though he reports severe stabbing and burning pains in his chest (possibly a heart attack?). Aaron progressed through a deep dark blue portal, traveled the universe, and passed the end of the universe into a creamy white light in the afterlife that he calls a "thing." Aaron tells us: "This thing I'm in, it must be God. It is not a man or a creature dictating over the world. It is just a force that is an endless and infinitely powerful "thing."... It is existence itself."[32] Aaron notes that the Energy did not call itself God, or a higher being, or any name at all. "But I am sure it was comparable to God. Although it was not a man and it was somehow one thing and thousands (maybe millions) of things all at once."[33] Interestingly, Aaron did not believe in God before he died. Now he calls himself agnostic.

Joan LH was twenty-eight when she had a spinal anesthetic as part of surgery to repair a weak tendon in her left ankle. Suddenly, the anesthetic moved up her trunk and stopped her heart and lungs. Joan, a Catholic, had a Bachelor of Arts degree in Theology. Her body was dead for 5 to 8 minutes. Upon encountering Source in the afterlife, Joan's beliefs were turned upside down. "God isn't an entity: it's a living force and it's LOVE.... My image of God transformed from the image of Trinity to a God that is truly All that is–a force, a power–LOVE."[34] Joan describes her belief system now as "new thought."

Ron Kruger's afterlife experience was what I call a meeting with a council of Light Beings who were monitoring his earthbound mission. I recognize the description in his account because I had two of these meetings myself. Fifteen-year-old Ron's human host died in an auto vs. apple tree accident resulting from his drunk friend speeding down a rural road that took an unexpected 90-degree turn. Ron flew through the windshield and became stuck between it and the rearview mirror, hanging by his head. Another passenger pulled him free without regard to his injuries, resulting in his body's death. In the afterlife, Ron encountered "an infinite expanse of glorious light ... [that] seemed to undulate gently with a force field."[35] Ron adds: "The ultimate spirit is an impartial force of universal and unconditional Love–a Higher Good ... the same force field of Supreme Love–which is also the basic substance of the universe."[36] Ron was a fundamentalist Christian when he entered the afterlife but claims after his return to have liberal beliefs.

Kathryn H's body died of a head injury she sustained when she nearly drowned in an irrigation ditch she was walking beside through a rice paddy in Bali. Kathryn tells us that in the afterlife she "had an awareness that a spiritual force exists that is all of us combined, not separate. If the word God is used, then God is all of us."[37] Kathryn was Buddhist before and after her expanded afterlife experience.

DW lost all hope after she was betrayed by a loved one. She lived alone in a small apartment in Nebraska and could not face either being alone forever or betrayed again. DW considered herself a Christian but did not attend a specific church. On a Friday night before New Year's Eve, in utter despair, DW wrote a suicide note and took a concoction she knew would kill her. In response to the question "did you gain information about the existence of God," DW wrote that she "knew the one Divine Loving Being I met was one who joyed in my creation. I believe there is a Creator."[38] She described this Creator as a huge golden, glowing orb being she saw in heaven. DW writes that her beliefs no longer fit within any religious denomination.

Demi B's fourteen-year-old body drowned in 1962. She went into a golden Light that she knew was where all things came from and to which they return. Demi was reared as a Protestant but did not believe that religion's dogma. When her body died, she was not met by relatives or any other being in the afterlife. Rather, she writes:

"I didn't see a God like being. I saw no beings but was immediately aware that there was a power there that has been described as God."[39] Demi knew instantly that everything came from, and returned to, this light/energy. Demi tells us in her description of the energy she experienced that she knew it was God, that it is what has been called God.

William W's body died as a child in 1985 when he drowned while visiting a California beach with his family. He was battered by large waves that kept him from coming to the surface to breathe. William describes himself as Christian-Protestant before he died and as having stopped going to church since then. He says he realized what his church taught was not accurate. William assures us: "Yes. God does exist. I was aware that God is not a man or woman. God is everything, a pure love energy" that he witnessed in the afterlife.[40]

Andy Petro drowned at a class picnic his senior year in high school. He had swum in the freezing water half way out to a floating platform to join his friends when he suffered intense cramps and went under. Andy had had eighteen years of Catholic education at the time of his death, none of which prepared him for the truths and reality of Source. Andy reports that while in heaven: "I knew the Light (God in Earth terms), because I became one with the Light."[41] Andy no longer affiliates with any religious group and calls himself spiritual and at one with Source.

Bobby R's body was four years old when he died during an acute asthma attack in a hospital emergency department. Bobby went into the Light and met what he believes is God. He remembers "a white light that seemed to pulsate and understanding that it was some kind of extension of God.... Somehow I felt the light from that device and God merge to become one with me."[42] Bobby was born legally blind but could see perfectly while in the afterlife. He was Christian at the time of death but now considers himself more of a Christian seeker.

Peter N's body died of medical shock five days before his twentieth birthday after a motorbike accident in Scotland. Peter crashed the bike when his passenger distracted him. Peter was trapped on the bike when it hit a parked car, rode it backwards as it rebounded, and landed on the road. Peter was agnostic to the point of being rebelliously anti-Christian before his death, though he was raised in a Christian home. Since then, he regards himself as spiritual. Peter met a Light of absolute love, personality, and intelligence. He writes in answer to an inquiry:

> that being of light that I was with I think many people would have no difficulty in saying that was God. I regard it as such. However, the difficulty I have with that is that the word God in no way comes anywhere near doing what it is any justice. It was way beyond any imaginings that I used to have in trying to conceive of what God meant.

Peter noted in his account that no religious teaching comes anywhere near what he experienced in the afterlife.

William Horden's human form died of a massive heart attack in a hospital emergency department on April 19, 2003. The body was dead for two minutes, but William lived on in the afterlife. William says the only true name he can use to describe the sphere of light he encountered there is the "Sphere of Universal Communion." "It appeared to me as a sphere of light, but light that is aware.... But with the additional sense of someone present," William reports. He adds: "the Sphere of Universal Communion is an infinite space of aware light that is occupied by all the individual spheres of aware light that ever have or ever will exist. As if it were One Mind."[43] William's Taoism-Zen-Native American beliefs did not change as a result of his experience in the afterlife.

Yazmine S's host died for six minutes after bleeding to death from a miscarriage in 1978. She describes meeting Source in the afterlife: "It was the One Mind. It contained the creation of all of everything ever created. I felt, I experienced everything that has ever been and ever shall be. All is simultaneously occurring. There is no past or future. It all just IS."[44] Yazmine describes her religious history as including bouts of going to the Church of England Sunday school, wanting to become a Christian nun, practicing Sufism and Mysticism, and ultimately adopting Tibetan Buddhism.

Shimmering silvery flow. Star or galaxy-like. A sea of colors. A source of energy. A force field of love. This is how expanded afterlife experiencers intimately know God the Creator. These expanded afterlife experiencers agree that God, the Creator, or Supreme Deity is a field or flow of Light, colors, Energy–a type of Energy that does not exist in the physical world. The Creator/Supreme Deity is the Energy source of all life and contains all life. It is for this reason that many expanded afterlife experiencers call it "Source" rather than God. "Instead of God, I will use the term Source, as it has neither masculine or feminine energy to distort the message. Source is the source, the ONE infinite consciousness," explains expanded afterlife experiencer Robyn.[45]

Each of the people quoted above suffered his/her body's death and entered the afterlife to witness the divine Source phenomenon in its true, raw, pure, spiritual and energetic form without any projection onto it of human concepts of what it should look like. We each report an afterlife experience that differs significantly from, and goes far beyond, those of the traditional near-death experience. I refer to us as *expanded afterlife experiencers* because of it. Almost everyone quoted in this book is an expanded afterlife experiencer.

But wait.

We have more to add to our description of the Source.

2

We Recognized Source's Collective Nature

EXPANDED AFTERLIFE EXPERIENCERS DID NOT see the Source as a being, much less as a human male, as religions predict God to be. Rather, we encountered Source in its raw, energetic form. That form is so completely foreign to human experience that no words to adequately describe it have ever been created. Therefore, we resort to weak analogies to earth life that, based on our individual life experiences, seem to come closest to approximating the concept. Because our life experiences and vocabulary are different, our word choices are different. But we are describing the same phenomenon: Source the Creator. There is much more we have to tell you about its nature. There is a crucial aspect of its nature that is even harder to clearly describe because it is impossible in human life.

Source Is a Multi-personality Phenomenon

More shocking to us, and probably to you, than Source's true nature and power is its multiple personality or collective constitution. Expanded afterlife experiencers, as distinguished from threshold experiencers, remembered, recognized, or learned that Source is composed of itself as a core identity and also of zillions of subpersonalities that we call souls. Experiencing this aspect of Source firsthand is very different than the common threshold afterlife experience of learning and feeling the unity and oneness in all of creation. Many threshold afterlife experiencers believe they have met God or are inside the unity or oneness of the Source phenomenon when they are flooded with overwhelming brilliant light and all-encompassing love. True, these are some of the ways that souls perceive Source's energy upon entering the afterlife. The love and light are certainly felt more intensely the closer a soul comes to Source's core. The unity or oneness experience, however, is merely the tip of the iceberg in terms of appreciating Source's mind and identity as being composed of different personalities in a way similar to how a novelist's mind includes all of his/her book characters, and our minds include all of our dream characters.

Andy Petro tells us in his YouTube video[46] that he witnessed a Source composed of an infinite number of pieces of Light that together constitute *the Light.* Anke Evertz agrees that Source is a lively Light full of souls that contains everything that exists. Doug F adds: "Everything is god. I would not use the word god" because it is too limited.[47]

Emanuele tried to go toward the beautiful colors he saw in the afterlife "but was blocked by a collective presence. I couldn't see the otherworldly Beings, but I could feel their presence. I instinctively knew that this collective presence was 'God.'"[48]

The collective or multi-personality nature of Source was also remarked upon by Kathryn H, who describes Source as a multi-personality spiritual force and an ocean of souls. "I had the awareness that a spiritual force exists that is all of us combined, not separate. If the word 'God' is used, then God is all of us."[49]

Wayne H understood, while in the Light, that all of the collective experiences and all of the collective knowledge of all parts of Source existed as separate items but also belonged to the whole.

> The whole was the collective knowledge of all. I understood there was no individual, just one, yet each experience was an individual making up the whole. This

> concept of ONE is so foreign to any description I can give, there seems to be no way now of describing it. My previous understanding of ONE was a single uniqueness. In this case, ONE is something else. Many being ONE and ONE being many, both existing simultaneously in the same time and space.[50]

Author Natalie Sudman, whose body was blown up by an IED in Iraq, describes her understanding of Source while in the afterlife as an all-encompassing reality:

> The *one* reality includes all *beingness* or consciousness. It is the endlessly unknowable infinity of creativity and an apparent paradox of infinite numbers of unique individuals that are simultaneously *one.* This encompassing connection is within and of, and creates, is created by, and moves through each unique being, and *is* part of all while also existing separately from what I'll call "All That Is."
>
> This All That Is can be perceived simultaneously as a force and as an individual consciousness that exists within each consciousness and yet is separate from each consciousness or being. It's what might be referred to as God, but the ideas of gods that we have are a pale and incomplete shadow of the All That Is that I perceive.[51]

Aaron M describes Source as "somehow one thing and thousands (maybe millions) of things all at once."[52] While Aaron was in what he calls the white place, he observed other, smaller "things" like Source forming something like a kaleidoscope of colors in mathematical shapes and patterns that floated all over the place, unfolded, grew, and glistened, or simply pulsated with energy. Similarly, Anonymous describes seeing pinpoints of light floating all over the place in the afterlife. He says they were souls that had different geometric patterns. Aaron's things and Anonymous's floating pinpoints of light are the Light Beings that newly arrived souls see in the afterlife.

During my first sojourn deep in the afterlife, I was met by my five closest, most beloved eternal friends in the form of Light Beings. The five glowed from within in unearthly hues differing from core to periphery and had brilliant auras fanning out like halos around them. My friends shared with me their joy at my homecoming, along with deep curiosity to know what human life had been like. One of them telepathically explained that they had rushed ahead of "the rest of us" because they were so excited to have me back. I understood at the time that "the rest of us" was Source. My friends were reminding me that Source is both a core phenomenon and a collective one. With my five loved ones in tow, I encountered what I can only describe as an Energy field that I call Source. It is both the source and the substance of all that exists. To my surprise, Source is nothing like a being, much less a human male. Nor is it a person in the way humans think of themselves as individual people. As I explain in my first book:

> In fact, Source is not an individual at all. God/Source is simultaneously **both** a single, unified *field* of energy **and** a collective being compose of all of creation, including us. In other words, there is only **one** being in our universe. All that exists in the universe, known and unknown to us, is part of the one being we have traditionally called God and I now call Source. All that exists constitutes One Being—Source.[53]

Souls Cycle In and Out of Source's Core

During the incarnation phase of eternal life, we souls cycle into and out of Source as we enter and return from living in the physical world. Using my simplistic analogy of Source being like our sun, these departing souls would be similar to coronal mass ejections of the sun's plasma. The difference, of course, is that souls may return to the Source between incarnations whereas coronal ejections do not return to the sun. Several expanded afterlife experiencers in my research group actually watched this cycling of soul incarnations happening!

Valeska K observed the Source to be a conscious, knowing energy source. She says many star lights orbited around it or entered and emerged from it in a logical pattern: "There was darkness with light star-like Beings and a huge source of light that they emanated to and from." She was there herself as conscious light. One of the other conscious lights told her: "This is home. This is consciousness. This is source."[54]

Henry W, while observing Source, became aware of what he calls small golden orbs of light and noted that though they were each separate individuals, "I knew that each and every one was connected and joined with what I believed was God."[55]

Bridget F also saw orbs of light going in and out of a vortex of light. "It was what I perceive to be the source. Just outside the hole were light orbs going in and out of the hole." She says they were different types of brightness, colors, and shapes.

DW, while observing the large golden orb she believed to be the Creator, saw "more 'glowing globe shapes' ... It was like a line of them approaching the largest sphere."[56] DW continues: "Now I saw a long oval of light with a pattern of tiny blocks in rows seeming to [be] moving along its length. A glowing golden light came off it like a sun and the love you could feel was like the large Loving Being sent out."[57] DW offers us another metaphor to explain the multiple or collective nature of the Creator as constantly changing, without itself moving, as souls cycle in and out of it for incarnation:

> Imagine a large, round, globe shaped zinnia. It's deep golden in the center and composed of many tiny petals. Starting at the center a small circle of golden petals appeared to come out from inside the being itself. There were four petals in this first circle. See each tiny petal as a moving, golden flame going outward from an ever-refilled center. …

> They were not expelled from it, like waste, but becoming, being created, from the power of the love within that Being. Creation as love made real, manifested.[58]

In human life, everything is considered to be "either/or"—a singular discrete experience. In the afterlife, everything is "and/and"—a multi-faceted experience that can have contradictory facets. My research subjects and I acknowledge the paradox that Source both is, and is not, a singular entity or power source. Source has two natures. It is the only consciousness in existence, the only mind, the only self-awareness, the Almighty. At the same time, Source is a collective entity consisting of zillions and zillions of souls and what humans believe to be individual creatures and things.

But wait. There is another important point to make about Source's raw, innate nature: Expanded afterlife experiencers discovered firsthand that we souls inside humans are parts of Source's collective nature—just like everything else in the universe.

3

We Learned We Are God

EXPANDED AFTERLIFE EXPERIENCERS NOT ONLY learned that the divine Source phenomenon has a unified nature as well as what we call a collective nature, we experienced ourselves awakening to being an intimate part of that collective. We did not merely feel united with all of Creation, or the sense of oneness that NDErs and threshold afterlife experiencers report. We each awakened to the eternal truth that we ourselves ARE Source—the sole consciousness and self-awareness that has temporarily taken on the illusion of individuality.

We Collectively Are God

Many of my research subjects learned through personal experience that God consists of everything in the universe and more. They knew themselves, and all of us souls, to be a literal part of Source.

Aaron M resisted accepting that he was in the afterlife, even though he felt deeply that he was home. A thousand voices within his mind confirmed to him that he was home, repeating over and over: "this is where you came from and this is where you'll end."[59] The voices urged him to let go of human life. Aaron protested that he was not ready to join them. Their response was that he was being irrational because he IS them already and it is time to come home.

Aaron could feel the white Light of Source's energy, what he calls "the whiteness," merging with him. "Suddenly the whiteness feels like it's fusing with me. It's so warm, it's flowing through me and I'm merging with it." The thousand voices told him: "This is energy, this is what everything in existence is made from, and that it flows constantly through everything." The voices explained that because this energy flows through everything, it means everything and everyone is connected. "We're all part of it." Aaron has a beautifully simple and eloquent explanation of this connection:

> The way I would explain it is like water. A fish is connected to all other fish in the ocean because the sea envelopes them all. They are all in the same water. That same theory applies to everything in existence. But instead of water it is energy. We are all made of and surrounded by energy. Even the air and gaps in space are filled with this same energy. We just can't see it. And this is the thing that connects us all.[60]

The voices told Aaron that he had now returned to the Source energy. When Aaron continued to protest being home, the voices reminded him to let go of himself as a person. They explained that his former body is just a form that Source energy has taken and now it is time to be reabsorbed into the whole.

"We are all pieces of the same Light." Andy Petro, a death by drowning survivor, realized he himself was a piece of the Light when he was in the afterlife. "Truth only exists in the Light," says Andy, and "the truth is that I'm a piece of the Light. And there are an infinite number of pieces of the Light who together create the Light. And that's reality."[61] Andy uses the words "the Light" for what others call God, Source, or the Creator. "We are all pieces of the same unconditionally loving Light."[62]

Andy remembers: "I was actually absorbed into the Light" and became an indistinguishable part of the Light.[63] Andy analogizes this sensation to a teaspoon of sugar being stirred into an infinite bowl of water, with him being the sugar and Source being the water. Andy explains that he was still himself, but he was also the Light.

Andy had known all through his afterlife visit that he was a piece of the Light, and that there were billions and billions of pieces comprising Source, but now he no longer had the perspective of being on the outside observing the collective entity. Like me, Andy experienced Source as having a great sense of humor and tells us that it is a fun place to be. Also like me, Andy states that nothing in his eighteen years of Catholic education helped explain what he had experienced. Andy jokingly calls his new awareness of his true identity "a heavy experience," one where you "realize that you are one with Light/God, actually an individual piece of God. Therefore, I am God, as we all are here on planet Earth."[64]

Anke Evertz, in a coma for nine days after accidentally setting herself on fire, reports in her YouTube video interview that she "eventually dissolved in the … what I call the Source today," and felt fully connected with everything and part of all. When her afterlife companion asked where she thought she was, Anke immediately replied "I'm at home" where we all are now and where we come from.[65]

Bobby R, only four when he died, calls Source "the Book of Fate" because it contains everything that has been, is, and ever will be. He remembers the Light from that Book of Fate and God merging with him to become one. Leonard completes this description with his report that the Light is Source and "we are not separate from this light, we are part of it, we are God!"[66]

Chantal L was the subject of a Code Blue in the hospital when a cyst the size of a grapefruit in her abdomen ruptured and she died of septic shock. She had an expanded afterlife experience while she was dead before the doctors were able to resuscitate her. Chantal informs us that all of her memories were restored in the afterlife as well as her true identity. "I no longer felt as a separate individual. I felt as if I was part of a collective consciousness. I sense billions and billions of beings and we were all One."[67]

DW, who died from a successful suicide attempt, tells us "we are each pieces of a greater whole as I understand it. And getting back where we can all be together again is the ultimate 'going home.'"[68]

While watching an oval of light from a golden orb/being, DW saw a pattern of tiny blocks in rows that seemed to move along its length. These are some of the Light Beings/souls in Source's collective phenomenon. DW noticed "This one was smaller. I asked that being what it was that was so pretty and so loving. It answered me. 'This is you.'"[69] This is how DW found out she is part of Source itself. After receiving many downloads of Knowings from Source, DW understood how intimately she was entwined with Source when it told her she had to return to her body, even though it might hurt to do so. She explains that "to hurt me would be to hurt itself in a literal way I can't explain well."[70]

Dea M's body was bleeding into her lungs, on life support, and eventually died in the ICU after a fatal rollover accident on a busy highway when Dea had her expanded afterlife experience. While in the afterlife, Dea could feel what it was like to be the infinite yet herself at the same time. She could feel her own soul touching that of the Creator and immediately "knew that 'God' does exist and we are a part of it all."[71]

Emanuele died of sudden cardiac arrest following a strong earthquake that rocked his house. After wandering around earth, in what he calls the astral plane, Emanuele decided to look toward the stars and saw a multicolored dazzling display of lights. When he tried to get to the lights, he was blocked by a collective presence that he knew instinctively was God. Emanuele could not understand how this was possible considering his lack of belief in God. In response, a telepathic message came: "You and I are one and the same. No one is separated from anyone. We are the beings that God created."[72] Emanuele learned: "in reality we are all the same Being and the same God, although in an individual form.... I saw that everyone of us is God. I didn't perceive divinity to be something exterior to ourselves."[73]

Joan LH, who was Catholic and had earned a college degree in theology before she died, agrees that each of us is a part of Source. "Each of US is a part of God, and God is far more than any one of us."[74] Joan learned that we are all interconnected as the One she knows as God. Joan describes what it feels like to be both an individual and part of a greater whole at the same time:

> We are all one. I did not lose my individuality and yet I was part of something tremendously more than just me. I could still feel "me," within the LOVING that we are all part of.... [W]e are part of God and we cannot ever be separated from God.[75]

Author Natalie Sudman agrees, noting in her book *Application of Impossible Things*: "It's recognized that we exist in concert, all experience being cooperative and parts of one source: All That Is."[76] Natalie was blown up in Iraq by an improvised explosive device (IED) that hit her police-escorted convoy traveling through the war zone. Fortunately, she was partially healed in the afterlife and sent back to us with her detailed messages of what she experienced before a council of beings monitoring her mission in human life.

Kathryn H, who died in a hospital from a head injury, admonishes: "What is most important is that we see that we are all part of one Being. We are not separate, but make things seem separate in order to play out this reality."[77] Kathryn herself joined the ocean of other souls in a state of merging their individual selves within Source.

Author William D. Horden paints a vivid portrait of Source's collective nature, which included William:

> The Sphere of Universal Communication is what I saw and what I felt.... It appeared to me as a sphere of light, but light that is aware.... An aware light that creates and sustains the possibility of shared awareness on a universal basis.
>
> I was fully awake [in the afterlife] when I realized I was myself a sphere of communion. I was a sphere of aware light: surrounded by an infinite number of other spheres of aware light. As I experienced it, then, the Sphere of Universal Communion is an infinite space of aware light that is occupied by all the individual spheres of aware light that ever have or ever will exist.

> As if it were One Mind, occupied by all the individual ideas it ever has, or ever will conceive; or a timeless, limitless, Over-soul, occupied by all the individual souls that ever have or ever will enter the realm of time, space, and personality.[78]

William W returned from the afterlife with the emphatic message that, "We are god because we are one entity."[79]

Yazmine S was filled with the love of the One Mind, which "contained the creation of all of everything ever created. I felt, I experienced everything that has ever been and ever shall be."[80] Yazmine calls this Energy the Great Presence, a clear and brighter Light within the Light that is her true goal, her real and true home, Source. Yazmine reveals that the Great Presence is what others call God. But she says no name can be given to that which is beyond naming. She tells us that she saw that "we are not separate; we are the entire One."

I Realized I Am God

A few expanded afterlife experiencers learned *everyone* is a *part* of Source plus they awoke to the humbling revelation that they themselves are the core identity of Source. Their sense of self-awareness expanded from that of an individual soul to the immensity of the All-mighty Source phenomenon. They tell us this is true for each of us.

Demi B, who was fourteen at the time of her drowning, writes simply that she understood very clearly that Source was the origin of all of everything, and it is that to which all will return. Demi answered a question of when she had the highest level of consciousness in the afterlife with: "When I felt myself expanding and realizing that I was returning to the source of all love and all wisdom."[81] Demi knew that the Source phenomenon consisted of everything and also realized she herself is Source. She writes: "I felt totally at one with this energy, it was a part of me, as if I was the energy source and was communicating with myself, but I knew it was a much higher power that I was a part of."[82]

Gwen J, who died from major bus accident injuries, reports that she learned in the afterlife: "We are one with God; we are God, just differentiated from God to get a physical body and come to Earth."[83]

Leonard, who survived death by heart attack adds to Gwen's statement: "We are the only true God. It's not that we represent one small part of the divine, but that we all are The Almighty, in its whole because this can NEVER be divided."[84]

Jennifer J reports awakening as Source during one of her four afterlife visits after being pronounced clinically dead from cardiac arrest, seizures, and a fatal stroke. She relates what it is like to be both an individual and Source at the same time:

> I was one with the Creator and with Creation itself. I was the Creator. We all were; those who haven't come back [to the afterlife] still are. It's impossible to describe.
>
> I was aware that my earthly body, my container or vessel of my soul had been shed, and I was so much more. I knew all things. I was God along with everyone else, and yet God was still there in superior existence, too: A universal power that was gentle and kind, humble and pure.... I had individual thought awareness of one being, yet was one of the whole, without definition or separation away from each other.[85]

Jennifer further explains the awakening phenomenon she experienced in the afterlife as:

> Yes, God or a supreme being exists; the irony of that is that we are all part of that one supreme being, which is our home where the heart of us originates and lives, but separated from it by our earthly physical vessels (bodies). God is supreme, so we are also supreme, but don't know it. Somehow fear separates us from embracing our true origin and existence.... It is our physical bodies that separate us from the One that we are, as water poured from a pitcher into individual glasses, where we stay until we die and return to the whole.[86]

Malinda K, who was dead for ten minutes due to a grand mal seizure, recalls actually being the energy she knows as Source. "And yes, I did say I WAS that energy, not that I only saw it and was drawn to it. I WAS it."[87] Malinda explains her change from Christian to atheist this way: "This is why I no longer believe in a separate omnipotent being called God. We ARE God. We are simply a beyond vastly intelligent source of energy."[88]

Mira Sai, formerly known as Arti Gupta, was filled with shimmering sparkling white-light energy and realized that there was no difference between herself and the Light Presence, which she says is the nature and substance of all of existence. "All are That One," Mira tells us. Mira could tell that she was not just inside the Love; she was one with it. And that felt completely natural to her. It is home. She affirms:

> It was true, there is only one Being, one God, and THAT is the true Self of all. All are just a reflection of God: All are That One. I, too, was That One. THAT is my reality. I was brimming, overflowing with the supreme knowing that The Supreme Being is my own true Self, my true identity.[89]

Mira Sai includes us all in Source: "you and I and all, are none other than that Universal Self: that we are all one, like waves in the Ocean, rising up for a short while and then merging back into the Ocean as the Ocean." She adds: "There is no 'other' … only One and that, all are That One."[90] She learned that Source, or the Supreme Being, or Supreme Consciousness is all that exists.

Pamela K, who died from a severe apnea attack, describes her awakening as Source: "I realized that in truth, the 'I Am' was who I really was and ALWAYS had been and always would be. What exultant bliss!"[91] Pamela's expanded afterlife experience account reveals that the experience totally changed her life and how she looks at the world. "Experientially I now KNOW that Love is all there is, manifesting in the limitless forms of creation—and there is no place in which God is not."

Robyn died from an anesthetic overdose and now sees God as something that she is a part of and inseparable from. It is not an individual entity full of power for humans to worship or serve, she advises us. "My true nature is one with all, and I am God. And so is everyone and everything else."[92]

Peter N, who visited the afterlife after dying in a motorcycle accident, describes his awakening within Source as follows:

> Then I became aware of the presence of a being of a power, magnitude and intelligence that was utterly indescribable and that was this light that I now knew to be here. (What I am trying to indicate here is that on first finding myself "inside the light" is that I "exploded" in size to unbelievable magnitude and that, in terms of having an identifiable "form" I just completely disappeared. Literally, I became one with what that light was, strange though it may sound it was as if I in some way became the light, I was completely merged with it.... It was like, briefly, becoming the light itself, and losing form because of that, then, once that part was over, reforming again but still being left in the light.)[93]

Sadhana, whose brief account in Dr. Rommer's book *Blessing in Disguise* reveals her metaphysical beliefs, explains that she knew to go into the Light and keep going as far as she could. She went beyond time, space, having any body, beyond the bliss of the afterlife, beyond experiencing, to what she calls the Godhead. "I went to the place where I no longer exist as a separate entity. It's like a drop in the ocean. You are totally dissolved. There is no separate consciousness," Sadhana tells Dr. Rommer.[94]

William Horden gives us a detailed description of his becoming aware that he was Source, using the names Spheres of Communion for souls and Sphere of Universal Communion for Source:

> I was fully awake when all the individual spheres of communion encountered one another at the same

> time, breaking through every dam of individuality and flooding us all in the totality of our shared being. This is why I suspect its name is the Sphere of Universal Communion, because when all the individual spheres of aware light periodically come into contact at the same time, every individual awareness, that ever has or ever will exist, is spontaneously and immediately At-One with the One.[95]

The effect, as William felt it, is like every drop of awareness in the ocean suddenly merging into a single awareness that it is ocean.

Yazmine S bled to death but returned to describe her realization that she is Source: "It was inside me. It was me. It was in and with everyone and everything."[96] She actually felt its energy pervade her very core, as if her entire existence was exploding into love.

On my afterlife journey through the corona-like Energy field to the core of Source, I was accompanied by my five most beloved eternal friends in Light Being form. In my mind, I visually compared what I felt as the Energy Field/Source with our Sun and its corona of gases and light. While well inside the afterlife, and after many exciting adventures, my five Light Being friends and I started sinking deeper and deeper into Source's Energy corona. We had no "beingness" about us. We were pure Energy and personality. I knew we intended to flow right into Source's core essence, which I understood to be one Energy field entity with zillions of discrete personalities within it. I could mentally and emotionally sense stronger and stronger Energy filling and thrilling me. Source's own memories started appearing in my mind the closer I got to its core.

Somewhere along the way, I lost awareness of my five eternal friends. I was alone but not lonely. Source's presence surrounded me with bliss and love.

All of the information I had been given when I first entered the Light, and during my traverse of Source's corona, started to gel in my mind. Suddenly, I was hit with the thoughts: "I am Source! I did this to myself!" I awakened to the knowledge that the amnesia of incarnation had hidden from me for so long—the realization that I am Source. I have never been separated. Never alone. Never left to fend for myself in human life. Never unloved. The realization that I am Source and had only been playing the role of the soul that inhabited Nanci filled me with such joy and humility. One would think that it would result in boastful feelings. But just the opposite was true. I was awed and humbled that little old me could be such a majestic divine phenomenon in real life. I was embarrassed by the realization that I had railed against so many events of Nanci's life, events that I had manifested myself as Source.

Instead of dissolving and disappearing, I—the personality I have always known as myself—expanded tremendously to encompass all of Source's mind, consciousness, and self-awareness. I could feel all the other personalities, the souls within Source. Like Peter N, I exploded into oneness with the entirety of Source. Like Andy Petro, Jennifer J, and other expanded afterlife experiencers, I retained my individual identity. Yet, at the same time, my mind and self-awareness expanded into that of all of Source.

Souls inside humans have stylized the God of their beliefs into humanoid forms because humans assume that a being of great power and intelligence must be like them. That construct locks the Creator into a "humans are separate from God" paradigm because humans are perceived to be separate from one another. The more expansive, non-humanoid understanding of the Creator as an Energy Source composed of all of everything, including us, that expanded afterlife experiencers gained opens up greater possibilities. It opens the door to answering how and why consciousness survives human death. Expanded afterlife experiencers were given those answers.

It all begins with Creation.

4

Creation of the Universe

THE SOURCE PHENOMENON'S DEFINING CHARacteristic, second only to unconditional love, is curiosity. Source is driven to explore anything and everything its infinite intellect can imagine. However, as Source informed me, because the intensity of its own energetic nature prevents it, Source cannot directly feel the more subtle sensations and emotions associated with everything it can conceptualize intellectually. It cannot feel traits that are opposite its own unconditionally loving nature. Nor can it experience physical sensations. Consequently, Source devised a brilliant plan for how to obtain firsthand, personal knowledge of what everything it could imagine feels like.

First, Source needed a less intense energetic environment within which it could have the exciting new experiences it craved. The solution was for Source to create physical matter and set it in motion to evolve into a vast universe of lower Energy environments. Malinda K explains how she felt as Source during Creation and its reason for creating the universe: "Just the complete feeling of curiosity and desire to experience the senses as a being. I felt myself to be highly intelligent and connected knowledge-wise to everything, yet is not enough. I wanted to be a physical part of that knowledge."[97]

Second, Leonard reports that Source's plan was to invest its own consciousness into the physical matter universe in order to experience it:

> Prior to universe creation there was only us, united in just one small point of awareness, this consciousness had knowledge, but we could not experience it, then we separated into billions of individual consciousnesses.[98]

Mira Sai describes Creation as divinity experiencing itself through its creations. She says that, just like herself, all of Creation, whether human, other animal, plant, or nature itself is Source—a glorious expression of its fullness.

Chantal L reveals the process she as Source used to create our solar system: "I remember distinctively how we created the solar system. I was part of this collective consciousness that had 'willed' it into being."[99] In response to a letter from Dr. Jeffrey Long, the founder of the Near-death Experience Research Foundation, Chantal explains more of her memories of Creation of Earth's solar system:

> I am sorry I cannot tell you exactly how we did it. I remember that it was our collective will and focus that brought it into being. Matter simply formed at our command (one minute it was energy and the next it was matter). It formed in the shape that we willed it to be. I remember how important equilibrium was. Every particle had to be in perfect balance and harmony with the rest.... There was a great joy among us, for this creation had a purpose of the highest importance. It would allow us to experience mortality.[100]

Chantal recalls creating only Earth's solar system and cautions that she does not know whether the same process was used for the rest of the universe.

Bridget F, however, did witness Creation of the entire universe from the beginning of everything and nothing. Bridget had to intentionally let go of some of this information in order to squeeze her mind back into her dead body. Similarly, Emanuele "understood that I and all the others on earth have created the universe as God."[101] He too cannot provide details. I can.

Near the end of my 1994 expanded afterlife experience, I relived Creation of the universe[102] in three different ways: (1) I watched with the same externalized perspective I held while inside Nanci's body, (2) I remembered being Source creating the universe, and (3) explanations were slipped into my mind whenever I was not certain what I was seeing. All three occurred contemporaneously thanks to our natural spiritual ability to hold multiple simultaneous levels of awareness.

I watched Source create a physical matter universe in its imagination where Source could indirectly experience the sensations and emotions it desired. Source used a process similar to, but far more complicated than, how we construct our dreams. What scientists call the Big Bang was Source concentrating and stretching its immeasurable Thought Energy into imagining a process whereby the entire physical matter universe evolved into existence from pure Source Energy. I learned that Source did not directly create suns, planets, or galaxies as finished products. Rather, I watched it expel[103] tremendous amounts of Thought Energy in waves and wavelengths that organized, reorganized, and gradually evolved into the physical heavens we see today.

Ron Kruger's body coded in the hospital from injuries received in an auto accident. Ron as soul, however, was in the afterlife watching Creation from his own viewpoint:

> Imagine, if you will, that this formless force was vastly infinite and evenly dispersed throughout infinity. Though it is perfect, singular, and whole, for the sake of clear rhetoric, I must describe it as having three properties. It is universal, unconditional, and benevolent. Being benevolent beyond our understanding caused the Force to desire other things to love, so it drew into itself with tremendous power and velocity, causing an

> extreme concentration of pure energy that caused an implosion, which fused energy into molecules that we know as "matter." In this respect, everything that exists is like a shattered piece of this Ultimate Force.[104]

Aaron M, an atheist before his human body died, similarly describes what he witnessed of Creation:

> I could see everything, not just our galaxy. I saw the whole universe. I could see the big bang. Although it wasn't a bang, it was more like a slow drip. It was like a material being stretched and slowly a hole formed in the middle of it. We are in that gap. I begin to understand it was this "god/energy" place turning itself into existence. It made a rule as if it made a game and set wheels in motion. Everything started from a small point and grew from there slowly. The universe is a thing that created itself and is watching itself. We're all part of it. This "place/thing/existence" that created us and is us, doesn't know where life will lead. That is what it enjoys.[105]

I too watched the Source phenomenon create and impose protocols, rules, or what we call laws of nature over its Thought Energy to guide its evolution. Those protocols include gravity, molecular cohesion, chemical bonding, the laws of physics, and all the other scientific principles humans have studied over the ages plus more. Bobby R, only four when his body died, was likewise told directly that Source put rules and limitations in place when it created the universe.

Source's Thought Energy eventually developed through chemical reactions and other laws of nature into biological matter as I both watched and remembered the process. From there, biomatter evolved into life forms. Through a series of intermediary species that I saw but do not recall, humans evolved into existence. Source thus "manifested" all of the creatures and things that make up the physical matter universe, including humans, through biological evolutionary and procreation processes. The word "manifesting" was used in the afterlife to name Source's ability to create what we perceive to be physical matter.

The entire universe, and everything in it, is nothing more than Source Thought Energy that resides firmly and forever within Source's mind and imagination. Leonard explains, "All that can be imagined may exist because it originates from universal consciousness (knowledge), thought is creative! Here things occur with a delay, but on the other side, they happen instantly."[106] There is no separation between the divine Source phenomenon and the universe it imagined. No planets or stars exist separate from Source. No human exists separate from Source. Although everything in the universe appears to go through a life cycle of birth, life, then death, none of that takes place outside of Source's mind. This is "how" life exists. Everything in the universe is Created within Source's mind and imagination, which divine power in and of itself gives life. Source created the perception of solid, physical matter so that it could experience all that it can imagine. But there is only one consciousness in existence no matter how many creatures there are in the universe. The Anonymous expanded afterlife experiencer sums it all up for us when he says that the Entity he calls Source created everything and remains connected to it all.

5

Creation of Souls

THERE WAS A SECOND STEP TO SOURCE'S PLAN TO obtain firsthand knowledge of all the sensations and feelings it could imagine existed beyond its innate nature. Source had to devise a way to enter into physical matter, without blowing it apart due to the intense force of its Energy, so that it could get the inside perspective.

Expanded afterlife experiencers know that there is only one consciousness in existence, that of Source. We also have personal experience that Source is composed of all of us. How, then, do we personalities within Source end up on Earth, or anywhere else in the universe?

We simply differentiate and incarnate!

I both watched and remembered Source designing its plan to experience everything it can imagine but cannot feel directly because of its vastness and energetic nature. The solution was to create individual small personalities/identities within its mind.[107] I call these mental characters "Light Beings" in my other books because that is what some NDErs call them when they appear in their afterlife experiences. For ease of reference, I simply call them "souls" in this book. Source's mental characters/souls are composed partly of character and personality traits of its own, such as creativity or a sense of humor, and partly of traits Source can imagine intellectually but not experience directly because they contradict its own innate nature. A couple of examples would be human greed and jealousy.

To be clear, Source's mental characters are not actually separate identities or beings. They are more like book characters a novelist creates in his mind or our own dream characters. It was explained to me that Source dedicates small parts of its own self-awareness and consciousness as these mental characters/souls, which in turn incarnate into physical matter in the universe.

My fellow expanded afterlife experiencers similarly learned some aspects of this plan. They refer to the creation of souls by the term differentiation, which is a much better fit.

When asked in the NDERF questionnaire accompanying her written account whether she had gained information about premortal existence during her afterlife experience, Gwen J replied: "Yes. I existed as God, then differentiated to become me in this lifetime. During the [afterlife] experience again I become one with God and had to differentiate again to become me once more."[108] In other words, Gwen learned that we souls float into and out of the collective entity Source for incarnation. That is how Source gets inside physical matter to experience it—via us souls incarnating into the physical universe.

Gwen elaborates on how it felt to rejoin Source after leaving her human incarnation:

> What I remembered was that I had completely merged again with God. ... I was no longer a separate being. I was where I belonged, where I came from. It was perfect. When it was time to return I had to again differentiate from God and become a separate soul again. Yet I was still a part of God.
>
> If our work is finished then we again merge with God. I had the sense that if our soul work isn't done that this merging doesn't happen. That was why other people have different near-death experiences. I had reached the ultimate goal.[109]

Atheist Sue C agrees with Gwen. She was told in the afterlife: "Everyone is connected to the source. We come from the source, we return to the source."[110] Emanuele, whose body died of cardiac arrest, briefly shares his birth as a soul in his report: "Before I was born, I existed as God the Omnipotent Father and then I was split from my true essence (God) to come to earth."[111]

Incarnation is an illusory in and out process. We souls never actually leave Source's mind because we are literally part of its own consciousness and self-awareness, which cannot be split. Peter N describes how, during his expanded afterlife experience, he reformed his individuality while also still merged into Source so that both states existed simultaneously:

> Once that "having no form of my own" ended it was like I re-formed into an identifiable form "out of the Light," as though I was extruded from it, though I was still in the light and still experienced myself to be, psychologically, massive in size, but I had "regained" an identifiable form. Sorry, I'm finding this extremely difficult to put in words. ... Then it came upon me that I knew I was inside this being and it inside me. We were merged so that there was no separation—and yet I also knew that I existed, as did it, as a discrete entity.[112]

Peter recognized the differentiation sensation and knew he had been in the afterlife before this incarnation.

I remember feeling what Peter N describes during my first expanded afterlife experience. To my great surprise, I was still me in the afterlife. I was still the personality I had previously thought was Nanci the human, complete with all the same likes and dislikes, the same sense of humor, and the same ways of thinking. Throughout my afterlife I still thought of myself as separate from Source—even though I had learned that it is a collective entity composed of all of everything. Then, after I had reached the core of the Source phenomenon, I suddenly awoke to the realization that I am Source and always have been. My sense of self exploded from Nanci's little frame of reference to the whole of Source's infinite All-mighty, All-knowing, Creator of the universe, Energy field vastness. Moments later, I contracted again into Nanci's unique personality just before I began a whirlwind journey back into her body. I had floated in and out of a sense of identity as Source in order to again differentiate myself into an incarnating soul. The switch was in state of awareness, not physical location.

DW watched souls being created in the heart of the zinnia-like Source phenomenon she describes seeing in the afterlife. She says the petal-like things were "becoming, being created, from the power of the love within that Being. Creation as love made real, manifested."[113] Source told DW that she was perfect because she was "As I made you. I did you perfectly!"[114] DW felt the love accept her completely just as she was.

Source imposes amnesia of our true nature upon its incarnating souls, much like we prevent our own dream characters from knowing they exist only in our imagination. This allows Source to experience genuine emotional responses to physical life from inside each discrete physical creature or thing into which a soul incarnates. This illusory separation into discrete beings or things creates the opportunity for Source to have imaginary relationships among its mental characters, like we have among our dream characters. When we dream, we create a dreamscape scene, populate the scene with people, and then, as the dream unfolds around us, we experience dream character relationships from the perspective of being inside one of the dream characters, usually the same person we are in waking life. The fact that we souls can dream inside our human hosts is evidence that we are Source. Our phenomenon of dreaming is a much watered-down mini version of Creation, and only Source can create physical reality. The biggest difference between dreams and Creation, besides the magnitude, is that Source invests its perspective into nearly all of the universe's physical matter at the same time, while we project ourselves into only one dream character version of ourselves at a time.

Expanded afterlife experiencer Henry W explains Source's purpose for imposing amnesia of our true nature on us souls:

> To successfully experience the human existence, one

> must be physically out of touch with God. We have to learn and seek on our own. We need to search out the meaning of our own existence and experience here on earth. Faith is the engine of discovery. Without faith, we are just like ants.[115]

This investiture of Source's self-awareness and consciousness into physical forms explains why consciousness survives death. Source's entire purpose for creating the universe is to be present within it and experience all that physical life has to offer. That could not be accomplished if small parts of its consciousness died after each incarnation, instead of leaving the body and uploading to Source all data collected during physical life.

You and I are those souls/Source's mental characters. "We are eternal beings!" exclaims Leonard. "We were not 'created' because we always existed, without beginning or end! We are God experiencing who we are through matter," Leonard tells us.[116] I learned in the afterlife that I, and by extension all of us inside human bodies, are not human animals at all. We are Source. Detailed understandings were provided to me in the afterlife to the effect that human animals are roughly as mentally sophisticated as other large animals. Humans have no more innate personality and character than a horse, dog, or elephant. Their primary behaviors of eating, drinking, eliminating, sheltering, sleeping, playing, and procreating are shared throughout the animal kingdom—as is violence. In the words of Aaron M, who calls Source "the universe" or "energy":

> The energy explained to me that my body was just a machine. My personality was just imaginary and shaped by the things I've seen. They said the only reason I exist is to see things and to experience life.... They also told me that humans are no different than animals, and that they were created equally. The universe wanted to create as many different experiences as it could, and that's why there are so many variations in life-forms and species.[117]

Our soul consciousness and personality survive the death of the human body because as Source we are eternal. I was informed through direct communication with the core of Source that there is only one consciousness in existence—the eternal consciousness of the divine Source phenomenon. I also learned that the personality, individuality, and consciousness we cherish as our individual identity is actually *not* that of the human body. It is the small part of Source's self-awareness inside the body that we call the soul. It is the soul that survives death of the body. We are the souls. All of the intelligence, sophistication, benevolence, love, creativity, and other wonderful aspects we have always called "humanity" are actually the traits of us souls living inside humans. Sometimes we gain enough control over our host bodies that our innate Source personalities are allowed to shine through their behaviors. When that happens, we convince ourselves that our bodies have those desirable traits and are not just animals.

I learned that Source consciousness does not reside in the brain, which is merely a computer-like organ that drives the biological bodily processes. Eternal consciousness resides in the soul. Soul energy is diffused throughout the body during gestation, and it withdraws and leaves the body at its death. But it never ceases to exist.

With this background on Creation, and who we really are, how and why soul consciousness survives death of its host human animal is obvious. The only consciousness that humans have belongs to Source. It cannot be extinguished. When the time comes for a human's life cycle to end, the same Source eternal consciousness that entered into the body before birth leaves at its death to continue eternal life.

6

The Purpose of Incarnation

WHY DID SOURCE CREATE SOULS TO INCARNATE? "Source has one desire, to know itself. Who am I? Source cannot know itself unless it divides itself into two, thus having another perspective to see itself from," reveals expanded afterlife experiencer Robyn, who uses the name Source because it has neither masculine nor feminine qualities to distort the message.[118] Sue C, another expanded afterlife experiencer, when asked the purpose of life responded, "We are all one. One is God. The separation of one to two is life."[119] It is that simple.

Source seeks to satisfy its curiosity about everything different from itself. Malinda K, while inside Source, "felt an overwhelmingly insatiable desire to experience things with all the senses I had held as a human. I needed to touch, hear, see, taste, learn, and experience emotion."[120] Source created the universe in order to be able to do just that.

Andy Petro learned in the afterlife that the pieces of the Light [Source] go down to Earth from time to time to experience "what can't be experienced in the Light because in the Light there is no time, there is no separation, there is no hierarchy. ... So, in order to experience evil, or happiness" the Light created a planet to go to in order to feel these things.[121] Andy explains that this was necessary because in the Light everything is a fantastic, euphoric existence.

Chantal L adds that Source wanted to experience the most dramatic difference between eternal existence and physical life. "I remembered [while in the afterlife] that we had [created the solar system] for the purpose of experiencing mortality. It was so important to my soul to come to earth and experience mortality."[122]

Robyn, whose body died from an anesthesia overdose, understood that the purpose of incarnated life is simply for experience and expansion. "We are meant to come here and play. Just be and play and experience this grand illusion of physical reality."[123] Leonard agrees, urging us to have fun here. "We created the universe to go there and have fun!" he says.[124]

In sum, Source's purpose for creating life was to see itself from a different perspective and achieve firsthand knowledge of everything different from its innate nature. It created us souls to be the small parts of Source consciousness to enter into the physical world to gather those experiences.

We souls carry out the grand design by incarnating into individual pieces of physical matter. Aaron M, who calls Source "the universe," outlines how all of this works. He understood in the afterlife that our own purpose for life is "simply to exist and to live."[125] He learned that the universe/energy/Source is always learning about itself and observing physical life through the eyes of its creations. Aaron tells us that Source likes to learn: "All we are meant to do is live and learn. There is no right and wrong. By simply being true to ourselves and living life, we are fulfilling our purpose," Aaron advises. [126]

Chantal L wrote to Dr. Jeffrey Long at NDERF that she remembers how incarnation works, phrasing it as: "There was a plan in place. We would take turns coming down into mortality and experience life in all of its aspects and bring this experience back to the collective."[127] "This is how it [Source] learns," explains the Anonymous expanded afterlife experiencer. "It does not just learn through observations, measuring, or repetition. It learns by being it and living it. And I understand that we are a very important part of this learning process."[128]

Joan LH, a trained Catholic pastoral counselor before her human life ended, laughed at religions while in the afterlife and said that we made life hard ourselves by adopting complicated meanings of life. She admonishes us that "it's really very easy. All we have to do is Be ourselves and express the Loving we are. We can do that by simply breathing in and out."[129] NDEr Linda G, who is not part of my research group, gives examples of how simple our life plans can be:

> Yes I came to understand that we all choose to come to Earth to fulfill a plan of some sort or even learn about a particular interest. We choose our bodies, parents, and life plan. May I also add that some people come here for the most simple of reasons: to learn how to play tennis for example, or simply for the cake and food - as silly as this sounds our life plans aren't so high and lofty as one might think. I DIDN'T MEET ANY "SAVE THE WORLDERS" OR ANYONE WISHING TO BE A PRESIDENT. MOST HAD SIMPLE WISHES.[130]

Some threshold afterlife experiencers interpret their experience to mean that the purpose of life is to learn human life lessons or to grow spiritually. Both assumptions are premised upon the common belief that we souls incarnate only into humans and that we remain human in the afterlife. As various chapters reveal, expanded afterlife experiencers received knowledge that these assumptions are incorrect and present far too narrow a view of life.

Expanded afterlife experiencer Gwen J reports that we souls are not incarnated here in order to learn human life lessons, as most religious and spiritual teachers claim. Rather, we are here to experience the joy of the physical world and fellowship with all things, which is what Source seeks.

Doug F sums up the purpose of life in the same way that Source explained it to me: "Another realization came to me. Everything is known and there is nothing to learn. It was clearly understood that there is no heaven or hell. Life on earth is not a school, there is nothing to learn. Life on earth is just an experience. Nothing more."[131] Andy M adds that the learning here has no purpose itself but is more in the nature of an experiment that the Creator conducts just to see what happens.

While learning important life lessons can be invaluable in this particular human life, the amnesia of incarnation will rob the soul of that wisdom if it reincarnates into another human. Chances are the soul will incarnate into some other creature or thing elsewhere in the universe where Earth life lessons are irrelevant. Human life lessons also have no value in spiritual life because we souls are no longer human in the afterlife. Expanded afterlife experiencers report that eternal life is nothing like physical life.

The nature of Source and of souls as my research subjects describe them demonstrate that it is not possible for the soul to grow to become more spiritual than it is. Souls are already Source itself, its consciousness and self-awareness. There is no higher state or level that can possibly be achieved. There is no need to attain a higher spiritual state or what many call a higher vibration. We souls are already there. Thus, incarnated life is not for the purpose of becoming more spiritual.

Three different goals were described to me in the afterlife as specific purposes or experiences souls could choose for incarnated life, in addition to the general goal of trying our best to overcome physical nature limits to acting with unconditional love in all we do. Each soul that incarnates chooses its own specific goals and purpose for coming into human (or any other) life. Natalie Sudman agrees: "each person is a unique being with their own purpose for coming into a physical body, and for creating meaning within that life."[132] The goals are not preordained or selected by Source or anyone else. They have nothing to do with what job one chooses, or spouse, or children, or any other aspect of life precious to humans.

The first, and most prevalent type of goal, is to experience a 360-degree perspective on a character or personality trait that interests the soul, a theme as it were—something unique to physical things or creatures and unlike Source's own innate nature. The type of learning or study involved when the soul chooses this purpose for incarnation is simply to experience all aspects of the trait in interest, to obtain experiential knowledge about that aspect of physical life.

Fulfilling this goal generally takes many, many incarnations all over the universe. When the soul is not actively observing or experiencing its chosen theme for study, it is free to do as Leonard suggests—to have fun and enjoy all that physical life has to offer.

For example, if a soul chooses to explore the human trait of jealousy, it may incarnate into what will turn out to be a jealous spouse or significant other and experience the trait from that perspective. In the next life, the soul may be the victim of other people's jealous behavior to understand how that feels so that it can relate those feelings back to what it felt like to be the one acting jealously. In another life, the soul may incarnate into someone who counsels jealous people who act out violently in order to get a third-party perspective on the trait. The soul will incarnate into physical matter creatures or things all over the universe that experience jealousy from one angle or another until that soul has observed and experienced the trait from 360 degrees of perspective. Some have misinterpreted this process to be karma because when a lifetime seems to be the opposite of a previous one, the conclusion drawn is that the new life is punishment for the former body's behavior. Reincarnation into an opposite lifestyle is not punishment, just a bold choice by the soul to feel the extremes of its chosen theme in serial incarnations.

A soul that has incarnated many times, and knows the ropes, may choose the second type of goal: to be present and active at a crucial moment in another soul's incarnated life in order to be a catalyst for what the other soul wants to experience. An example would be someone who appears in a person's life—such as a teacher, friend, or doctor—just when they are needed most to change the direction of the life of the soul needing a catalyst. The only spiritual purpose, if this is the goal, will be to perform the needed task. The rest of the catalyst soul's incarnated life may be devoted to pursuing fun and other physical interests.

Other souls experienced at incarnating may choose to incarnate with another beloved soul in order to be that soul's support through physical life, either because the other soul is inexperienced in that type of incarnation or because the life it has chosen will be exceedingly difficult. The soul choosing to be a support person will incarnate into the other soul's spouse, child, parent, friend, religious confidant, teacher, or other person to whom a person would normally turn for help or guidance. This purpose often requires years of contact with the other incarnated soul in order to provide the support. When not providing support, the soul is free to pursue its own desires.

Knowings informed me that when the soul is not experiencing events directly related to its specific goal, the rest of the soul's life has no spiritual purpose at all. But each soul's adventures in physical life are dramatically important. This soul work is what makes Source's data downloads feel like firsthand experience and not just intellectual knowledge. Everything we souls experience and perceive in the physical world gets added to Source's vast range of knowledge. It is part of what makes Source All-knowing.

Additional Incarnation Information

I learned in the afterlife that while part of our spiritual self incarnates, another part stays in the afterlife. Andy Petro mentions in his YouTube video that he was given the information that we souls leave part of ourselves in the Light/afterlife when we incarnate, resulting in our bilocation. Chantal L also reports that a small part of her took physical form to play the role of the human named Chantal. Afterlife data downloads informed me that the part of us that stays in the Light when part of us incarnates as a soul can act like a type of guardian angel helping the soul navigate its purpose without disclosing that purpose to the incarnated part. Henry W explains why our purpose is not disclosed to the incarnated soul: "To successfully experience the human existence, one must be physically out of touch with God. We have to learn and seek on our own. We need to search out the meaning of our own existence and experience here on earth."[133]

While in the incarnation phase of eternal life, souls residing in the afterlife have access to all of Source's wisdom and understanding with a few exceptions. They cannot access the truth about who and what they really are or the fact that their existence rests solely within

Source's mind and imagination. Afterlife inhabitants still believe they are separate beings from Source, just like souls inside humans do. They think they are actual beings rather than minds existing at a particular level of awareness. It is this fact that makes asking them for advice dicey. I know from my times in the afterlife that no soul in the afterlife that is still in the incarnation phase of life truly understands the information it can access within Source's mind until it completes later phases of eternal life.

When I was in the afterlife the first time, I received a lot of information about incarnation. Unfortunately, I now recall only a few facts. I remember that incarnation is entirely voluntary. Doug F notes the same thing in his expanded afterlife experience account. Each soul decides if, when, how often, and into what physical matter it will incarnate.

Incarnation is not a punishment for things done wrong in a different physical lifetime. Nor is it part of any mythical karmic cycle. We choose our own incarnations. As Jennifer J reports: "We choose life to start with, and we can choose reincarnation into this world or another world, this universe or dimension or another, and this form or a different one."[134]

Jennifer and Aaron M affirm what I was astonished to learn in the afterlife: we souls incarnate not just into humans but into any type of physical matter we choose, at any location in the universe.

Jennifer dispels the common understanding of incarnation with this explanation: "Everything is a soul and spirit, even the rocks and the hills.... I realized especially the significance of the nature of being a soul with a body, rather than the other way around, which is the way that western religions teach it."[135]

I was surprised to learn that we choose the parents of the human into which we desire to incarnate—not the specific child itself. Parents are chosen on the basis of their genetics and lifestyle because those factor into whether a particular soul will be able to meet its intended goals. I also witnessed that we preview some of the events of the lives of our selected parents' offspring in order to determine whether one of those lives will suit our purposes. After we incarnate, when one of those previewed events comes to pass we souls experience déjà vu.

Aaron M relates that our time in the afterlife is temporary while we are in the incarnation phase because Energy has to be reborn again and again. He also learned that he would transform into something besides a human someday. While they were in the afterlife, many of my research subjects were given access to their memories of other physical lives they had lived.

Valeska K, who received many flashes of information about the purpose of life, comments that it is difficult to explain. She sums it up with something we all know intuitively: "It is as if the journey is in itself the meaning, or a lot of it."[136]

7

No Bodies in the Afterlife

Many spiritual and religious people assume that we souls in heaven have bodies that resemble a spiritualized version of the human form. They call this phenomenon an astral body or ephemeral body. Religions also seem to assume that we will continue to be human in heaven, some going so far as to predict the resurrection of decomposed bodies to join their spiritual counterparts in heaven. As a result of these belief systems, many departing souls expect themselves to continue to be human in the afterlife. They find out pretty quickly that this is not true.

Near-death experiencers advise that we have no physical-like body after leaving the human one, reports psychologist Kenneth Ring, PhD, a renowned NDE researcher. His research finding on this topic is:

> When individuals claim to be out of their body, are they typically aware of having "another body?" The answer is No. Most persons are simply aware of the scene before (or below) them. When asked about "another body," they usually respond that they are unaware of having one or that they felt they existed, in effect, as "mind only."[137]

Expanded afterlife experiencers agree that their human body is gone. They provide a variety of reports on what form they took in the afterlife.

No Human-like Body

Some of my research subjects did not mention one way or another what they looked like in the afterlife. But a few specifically stated that they did not have a human body. For example, Aaron M writes: "I notice that I don't seem to have a body anymore."[138] Likewise, Dea M, who died during a rollover highway accident, did not have a body in the afterlife.

Jennifer J had four dying experiences, two of which happened when her body coded and was pronounced dead in the hospital. During one of these episodes, Jennifer noted: "I had no physicality, no shape, and no form. I still had thought but with no eyes or body."[139] Likewise, during one of her NDEs, Jennifer existed only as perfectly lucid thought. In her expanded afterlife experience report, Jennifer notes she "was aware that my earthly body, my container or vessel of my soul had been shed, and I was so much more."[140]

DW admits that she was still thinking and feeling in the afterlife, but not from within her body. "I was moving without legs. I couldn't see myself so I assumed I had no legs, no hands, no arms, no feet or anything you would normally look at to see if it was there."[141]

Kathryn H reports being light and bodiless in the afterlife.

Sadhana describes her time in the Light as one of going beyond and beyond, far beyond the point where you have any body whatsoever. After that, you have only awareness and a feeling of freedom, she tells us.

Leonard admits he was curious to know what he looked like without his human body. "Then I looked at myself and I was light," he writes, "I was made of light!" Mira Sai also reports that, while in the Light, "I looked at myself for the first time and saw I had no body. I was just a spark of Light."[142] Valeska K agrees that while in the afterlife she was made of conscious light.

William Horden entered what he calls the Sphere of Universal

Communion, a sphere of aware Light and his name for Source. He realized that he too was "a sphere of communion. I was a sphere of aware light" in a bodiless state within Source.[143] William adds: "Although it is much more difficult to perceive here, than in the Sphere of Universal Communion, we are no less individual spheres of communion here than we are there."[144]

We may change our perception of ourselves as we move deeper into the afterlife. For example, Leonard, after having first observed himself as Light, later felt that he no longer had the light form. He realized, "I was just a point of consciousness in the universe!"[145] I had the very same experience. When I first entered the Light, I seemed to be some type of vague invisible form made of Light. Later, as I moved deeper within Source, I became aware that I was just a mind, or a point of consciousness. Pamela K similarly felt only the "I" part of her floating freely in the darkness of deep space. Peter N reports that he seemed not to have a body in the afterlife. He existed only as consciousness.

Ron Kruger describes leaving his lump of meat body behind when he died and being in the afterlife without it. Ron learned a very surprising thing about physical appearances, advising us:

> A lifeless body without a soul has little distinctiveness. In fact, most of the distinctions we notice in the faces and body shapes of our fellow men are largely exaggerations of our minds. They are the ego's habit of isolating us from our fellows and of judging others based upon appearances. When we die and realize a universal connection to all mankind through the same life force, these distinctive features blend and blur into a general shape and look of man.[146]

This may account for why some near-death and afterlife experiencers, like me, did not recognize their bodies once out of them.

A couple of expanded afterlife experiencers give us more detail about their former physical functions as transformed in the afterlife. For example, Henry W tells us that he could see 360 degrees as soon as he was out of his body. He could also see colors and hear silence. "I was breathing but not, instead of air I felt a living force flowing through me. I felt as though I was swimming in the very Essences of Love."[147] These sensations are the only aspects of the familiar human form that Henry mentions.

The first thing I noticed when I entered the Light was that all of my pain was gone. I had been in physical pain for a decade from neck injuries and simply being a gym rat. While initially in the Light, I used the physician's CPT code guidelines[148] for conducting a physical exam of myself in an effort to diagnose my new medical condition outside my body. I had no idea that my body had died because I never lost a moment of consciousness. I mistook my heart stopping for just my heart calming down and not beating as hard as it had been. In the Light, I found that I had no heartbeat, no breathing, and nothing more than a vague sense of physical form. Still, I did not think I was dead because I was clearly alive and more aware and conscious than I had ever been. I could see in all directions, including through the back of my "head" via invisible eyes that seemed to be located about where they are in a human head. There was nothing to see except Light everywhere. I could not hear anything. I was under the impression that there was nothing in the Light to hear because I was alone. I could not smell anything. There was nothing to touch inasmuch as I had no body and there was no physical environment. Later, once I was deep in the afterlife, even the sensation of any kind of form vanished. Like Leonard and others, I felt myself to be simply a point of consciousness with memory, intelligence, personality, and emotions.

Only one expanded afterlife experiencer describes having a body-like shape while in the afterlife. Bridget F refers to herself as having an astral body once out of physical body but does not describe it as having any human features.

But I Was Still Me!

Do not be concerned by the absence of your earthly body in the afterlife, Sue C wants us to know, because we are still ourselves there. She reports: "I can assure you that after death you will feel completely as you do now, whole, even without a body."[149] I can vouch for that conclusion. So can threshold afterlife experiencer Beatrice W, who died from anaphylaxis after a restaurant meal. She reports her observations of herself in the afterlife: "even though I couldn't see my body anymore, I was still 'myself' with all the feelings, character traits and thinking."[150]

Some expanded afterlife experiencers report seeing themselves as part of the Light after they entered the afterlife. Andy Petro confirms: "I was still Andy. I was everywhere and I was here at the same time. I saw me as a person and I saw me in the infinite warm and loving Light."[151] Andy never describes himself as having any kind of body.

DW says she was confused at first in the afterlife because she could not understand what was happening to her. She reports that "'I' was still 'me.' I was, apparently, alive. I could not see myself. I could not raise my hand to look at it, but I was something—I still felt like 'myself.'"[152]

Joan LH explains that in the afterlife we remain ourselves while we also become part of the Light: "Individuals [in the afterlife] did not exist in the same way as we do here. I was still me, but I was also part of The Loving [her term for Source]."[153] The Anonymous expanded afterlife experiencer notes at one point in his narrative: "I have been liberated from my body. But, I still feel like I am me."[154] He states he exists only as soul and consciousness in the afterlife. Anonymous describes soul as "pure Energy connected with consciousness and intelligence."[155]

Robyn saw her body from the ceiling of the surgical suite and went home, into the afterlife. She reports becoming one with all existence, yet she had a firm knowing that she was still herself.

Ron reports having his true identity intact in the afterlife, without any body. He says his true thoughts and emotions, his true conscience, survived without sensory input or the influence of the body's survival instincts.

Based on three expanded afterlife reports with which I am familiar (including mine), I surmise that it is possible to continue to have a spiritualized human body in the afterlife if one is meeting with the council of Light Beings monitoring the soul's mission. This is a very different type of afterlife experience than one has upon dying at the end of life, and it has many different features.

Natalie Sudman had what I call a council meeting afterlife experience, which occurs when a soul is engaged in an ongoing mission and is called back to the afterlife to report mission progress. As far as I can tell, only souls with specific missions for the benefit of all of the collective Source meet with a council of Light Beings monitoring their progress. Perhaps that distinction accounts for why Natalie perceived herself in human form in the afterlife. She saw herself standing in her bloody and torn fatigues, her body dirty and well-tanned, addressing the thousands of Light Beings on her council. Natalie had been blown up by an IED in Iraq and went straight from death to this council meeting. More detail is provided in chapter 15, "Light Being Council Meetings."

I have had two council meetings, described in detail in chapter 15. During the first one, I had no awareness of whether I had a human form or not. Before the second one began, however, I perceived myself walking into the meeting space and sitting down in a chair. For just a couple of seconds, Nanci's parents appeared to join the meeting wearing the faces I loved when they were incarnated. Then they resumed their vague Light Being form. Yet another familiar soul joined the group in full human body physical matter appearance. I watched him lift a very human looking leg onto the edge of the table and speak to me moving human lips.

With the exception of council meetings, which are specifically designed to be brief, temporary visits to the afterlife, none of us expanded afterlife experiencers who turned our attention to the issue discovered any evidence that we continued to be a physical human being in the afterlife. Rather, we reverted to our true spiritual nature.

8

Expanded Life Reviews

READING AND HEARING A COUPLE THOUSAND NDE accounts revealed to me two different types of life reviews, happening at different stages of the afterlife. One fits the old adage that, when you die, your whole life flashes before your eyes. This type of life review can happen before or after death and can consist of all of the soul's current incarnated life's events in order, or a random selection of them. For example, NDEr Greg Nome nearly drowned in the churn of a waterfall and objectively watched a rapid review of scenes from his childhood that he had forgotten, as well as some scenes from young adulthood.[156] These events may take the form of memories or could appear as scenes on a movie screen or pages in a book. NDEr Norma E, for instance, watched her former human life on something like a huge TV screen divided into thirds. The left third of the screen showed the events of her life. The middle third showed what could have been. The right third had nothing but check boxes labeled "goal not achieved."[157]

The second type of life review usually occurs deep in the afterlife and constitutes part of the automatic transition from thinking like a human to returning to our natural spiritual state of mind. During the deeper life review, the soul sees all the events from the just past human life in a different light and much more.

Scope of Expanded Life Review

Knowings I received explained that one purpose of the more comprehensive life review that expanded afterlife experiencers have is to answer all of our lingering questions about why things happened the way they did in human life. We get to hear other people's thoughts as well as our own, so we can understand their perspective on events. We feel their emotions in response to our actions. We come to know others as we know ourselves by seeing into and feeling their hearts and minds. We are also gifted with the vast perspective necessary to understand why our choice in a particular instance was the best or not the best for us. The inclusion in Norma E's life review of what might have been is part of this new perspective.

Leonard briefly describes the nature of the expanded life review: "God then showed me all my life from birth until NDE. I felt and experienced again all these events and I also felt emotions I had raised in other[s]."[158] Leonard's life review is representative of those my research subjects had.

For example, Anke Evertz, who accidentally set herself on fire, suffered third degree burns, and died while in an induced coma for nine days, confirms Leonard's report. Anke describes her life review as an event where she not only relived and felt everything from her just lived human life, but she also felt others' thoughts and feelings and understood their viewpoints and motivations. A Light Being accompanying her explained the connections between and among the events of her human lifetime. As in Norma's case, Anke was also shown how things would have turned out had she made different decisions.[159]

Yazmine S saw and reexperienced every detail of her just ended 35-year long human life, like watching a movie and having the starring role in it at the same time. She realized watching this that the pain, fear, and earthly conflicts are unimportant once you achieve the state of ultimate freedom and blissful awareness she enjoyed in the afterlife.

In my first life review, the scenes of Nanci's life were displayed swirling all around me, as if being projected in no particular order onto something like the surface of a giant soap bubble. One of my eternal Light Being friends who greeted me in the afterlife was inside the sphere of memories with me. Drowning victim Andy Petro saw his life displayed on a huge bubble as well. He was also accompanied by a Light Being during his life review. While my review of Nanci's life was playing out, my four other Light Being friends were popping into various scenes from my life and living them vicariously. I had no idea before then that such a thing was even possible. This phenomenon is discussed in chapter 13, "Living Other Souls' Lives Vicariously."

I attended my second life review alone. I had it during a brief return to the afterlife when I died from unknown causes about five years after my 1994 expanded afterlife experience. The review was current to that date and included my body's future. The detail was phenomenal. I experienced not only my own thoughts and actions, but those of the people with whom I was interacting. For example, I watched a meeting I had attended with several attorneys in my old law firm. A young attorney I was mentoring sat across the conference room table from me. During the life review, I saw myself kick his leg under the table. He felt that I was rebuking him for something he had said. I was actually just clumsily crossing my legs. I also saw all the times that I had unknowingly physically hurt someone by patting them on the back when they had a sunburn. I felt their pain and consternation in the life review. During this second life review, I understood that our intentions are as important as our actions.

Another person in my research group also had a life review after she had returned to her body from an expanded afterlife experience. While objectively watching her body in the ICU, Dea M saw a vision of her whole life projected upon a hospital wall. The vision included events with her parents, friends, and family and showed Dea how all those experiences were woven together. The vision was so intensely emotive that Dea felt as though the emotions it generated were drawn right into her body's chest. Dea understood from watching this review that her life was exactly what it was supposed to be.

During both of my life reviews, I watched a replay of my every thought, word, deed, and physical data input, and saw and felt the emotional impact of my body's actions upon not only me but also on everyone around me. Andy Petro describes the same phenomenon. I learned that we souls have an innate trait of total recall of everything we perceive and experience, so an in-depth life review is quite a show. Carol I, who did not have an expanded afterlife experience but did have the expanded version of the life review, describes the multilevel presentation perfectly:

> You could call it a "life review" but it was more in-depth than that. It was multi-faceted. These were the facets. First, I experienced incidents from my life from my own point of view, second, from the point of view of whoever was with me, and third, from the point of view of a witness, a watcher of sorts, all simultaneously.[160]

What Carol describes is possible in the afterlife, but not in human life, because as parts of Source we have its ability to hold multiple simultaneous levels of awareness. I experimented with this ability early in my first afterlife adventure because I found it so fascinating. I kept expanding my perception and awareness of myself outward in concentric rings while at the same time keeping all of the awareness from inner rings intact.

Bobby R's deep life review demonstrates that the age of the human body the soul leaves behind has no impact on what the soul experiences in the life review.

> I saw my life, all 128 years of it. ... I was shown the consequences of my life, thousands of people that I'd interacted with and felt what they felt about me, saw their life and how I had impacted them. Next I saw the consequences of my life and the influence of my actions.[161]

The life Bobby reviewed spanned 128 years, not the four he had lived prior to his body's death. It also included previews of events that were to take place in his parents' and sister's lives. He watched his own future as a writer and witnessed his beloved body's final death. Those visions haunted the four-year-old after he returned to Earth. He feared that every illness his mother suffered would lead to her death as he had previewed it in his life review. He felt guilty about not being able to stop his sister's future rape. A Light Being had forewarned him that his life review would cause him pain because he would remember it once back in the body.

Sue C describes a unique life review presentation. The time line of her former human life was running in a figure 8 lying on its side (the infinity symbol). The clips of scenes included not only her conception and birth, but also her future elder years. She got the distinct impression that the entire thing had played out millions of times before. She did not specify in her account whether she meant the life review phenomenon or the unique events of her lifetime as Sue C had played out repeatedly. Sue felt at the time that she could pick any scene from her human life to return to upon surviving death. Nevertheless, she ended up right back in the hospital being resuscitated.

Other expanded afterlife experiencers, like Gwen J and me, also saw the future of the human body to which we would return. Emanuele saw scenes from his body's future during his life review. Gwen cannot recall that future now but remembers seeing it.

Self-evaluation During Life Review

Dr. Bruce Greyson reports that a quarter of the NDErs in his thousand-plus research sample had a life review. Of them, half believed the review is a form of self-judgment. They also believed religious doctrine to the effect that what happens to souls in the afterlife partly depends upon how their human bodies behaved, in terms of its good and bad actions. From the perspective of my research, and having progressed further into the afterlife than NDErs, the belief about human behavior controlling afterlife events reflects purely human misunderstanding. The NDErs in Dr. Greyson's sample projected their religious beliefs onto their NDE because they did not progress far enough into the afterlife to gain any other context.

Some in my research group also felt that they judged their human lives as they relived events in their life reviews.

Anke Evertz reports that there was no judgment by God or others during her life review. She did, however, judge or evaluate her own actions, though she cautions that "judge" is not the correct word to describe how deeply moved she was while watching how badly she had treated herself.[162] Leonard also related that during his life review "I was my only judge! … This is what hit me the most. God does not judge, he just loves us with unconditional love."[163]

Bridget F saw her human life from an objective perspective, which she assumed to be Source's. She saw the effects of all of her selfish acts as well as her manipulations of others. The review caused her significant emotional pain inasmuch as she never considered herself to be a bad person. A Light Being came to her then and told her that she acted merely as a human acts and that there is no fault in being human. She felt forgiven with those words.

Yazmine S was her own judge and felt ashamed that she had not lived her human life in a state of serene joy. She felt as though she had let down The Great Presence, which is what she calls Source. Yet, she simultaneously felt great love during the review.

No Judgment or Punishment

Other expanded afterlife experiencers' accounts confirm that our bodies' actions are relived in the afterlife, but never judged or punished by Source.

In the words of Andy Petro: "The Light also knew everything that I have ever done or will do, and the Light loved me without conditions. There was no fear. No judgment. No punishment. No blame. No shame. No ledger of good and bad deeds."[164] DW concurs, noting that "That being knew all of everything I ever was and loved me. Not just loved me but everything that defined me as myself … It loved the way I was made. … I was perfectly what I was supposed to be and it loved me just that way."[165]

A data download I received in the afterlife informed me that the purpose of the life review is not to judge the soul, nor to punish it, for what its human host did. The review is part of the transition process that leads the soul away from thinking like a human, with its ideas of revenge and punishment, and back to thinking and feeling like an unconditionally loving Light Being. We are able to watch, and feel, all of the events of our recently departed human life, as well as all of the hurtful things our host bodies did to others, from a new perspective of nonjudgmental love.

Bobby R states he felt no judgment during his life review. Instead, he felt pride, love, joy, and sometimes sadness as he watched from the perspective of a third party looking on events with him.

Chantal L recaptured all of her memories from her human life as sudden images in her mind. She reports that she did not have to account for her human acts and was loved and cherished no matter what she had done. Her only pain was that she felt she had failed herself because she had not completed her chosen mission.

Pamela K had a life review immediately upon entering what seemed like the darkness of deep space, where she found herself floating freely after her death. She turned to look back at her life and saw everything happening simultaneously. At first she felt apprehension, but then she realized that she was not going to be judged by anyone but herself. Pamela writes: "God does not mete out 'justice'—he doesn't have to. WE are the ones who allow ourselves to return to a heavenly state when we've learned to embrace everything in life as a manifestation of God, including our very own hearts."[166]

Natalie Sudman, whose afterlife experience was a meeting with Light Beings monitoring her mission on Earth, nevertheless writes in *Application of Impossible Things: My Near Death Experience in Iraq* that she encountered no hierarchy of power or judges while she was in the afterlife, even among those monitoring her human life. She assures us that there was no third-party evaluation, much less punishing judgment. She did not even self-evaluate her former human host's actions.

Peter N gives a beautiful description of the type of "judgment" that occurs during a life review in the afterlife: "(When I say 'judgment' I would very much underline that there was no prospect of any kind of condemnation involved in this.) I felt so at ease, so nurtured, so wrapped and rapt in the care and concern of this [Light] being for me that I knew it would never do anything to harm me."[167] Peter felt no concern about the life review and was happy to have the being look into him with unconditional love. He sums the review up as the being's search for the answer to the question of "What is, was, the essence of your love?"[168]

Ron Kruger, whose afterlife experience also included a meeting before the Council of Light Beings monitoring his earthly mission, describes the life review as an event where we can glean the maximum benefit from our earthly existence. He revisited scenes from his current human life and felt the actual pain or anguish, pleasure or love, that he had caused others as though he was the recipient. Ron states that the purpose of the review is not punishment, but, rather, spiritual growth through understanding the ramifications of our actions while incarnated, which increases our compassion for others. "The ultimate irony," writes Ron, "is that every time we hurt someone else, we eventually hurt ourselves."[169]

An even more insightful and wondrous perspective was available to those of us who remembered other physical lives—what most call "past lives"—during our life reviews.

9

Memories of Other Physical Lives

ANOTHER HALLMARK OF THE EXPANDED AFTER-life experience account is that it gives us a peek into the eternal lives of the souls who have gone this deep into the afterlife. These reports make it clear that the term "afterlife" is a human-centered concept that conceals the expansiveness and infinity of eternal life. The term assumes human life is the epitome of existence and heaven comes second. In truth, life is eternal because the font of life, Source, is eternal.

Our first glimpse of eternal life comes with the unfolding of expanded afterlife experiencers' memories from dozens, hundreds, and even thousands of other physical lifetimes they have lived in addition to the human one just relinquished.

Native American-Taoist William Horden refers to us souls in the afterlife as having memories of "the accumulated impressions of all the lifetimes we recall, the sum of all the personalities we have yoked to our soul, our enduring storehouse of mortal treasures."[170] In less eloquent words, this one human life is not all there is. Threshold afterlife experiencers who enter the afterlife and have a life review often see events from only the human life they just left. But, some expanded afterlife experiencers' life reviews include what we call past lives, one of the many events that distinguish an expanded afterlife visit from the traditional NDE model.

My former Catholic religion denied that reincarnation is real. That is what I believed at the time of my first death. The truth of reincarnation was proven to me in the afterlife when I suddenly remembered all of the hundreds, and possibly thousands, of other physical lifetimes I had lived. These memories swamped me at the same time that my review of Nanci's life was playing out on a bubble around me. I lost interest in watching it, thinking, *been there, done that.* Something far more interesting was happening. I recaptured all of my memories of life as amazing creatures and alien things far from Earth's orbit. These memories downloaded into my mind as a unified whole. I could recall every single moment—every event, thought, hope, dream, and sensation of each lifetime. I felt complete without realizing I had always felt otherwise. I was finally wholly myself again.

I remembered and relived scenes where I was various types of animals, a large variety of inanimate objects and plants, what we would call alien beings, gaseous clouds, and other matter throughout not only our galaxy but also the entire physical universe. The most astonishing past life I remember was when I lived as a nebula. I could feel myself scattered in something like droplets over a large area of space. I could also feel the empty space in between the droplets, but they did not interrupt the continuity of my existence, consciousness, or thinking. It was exhilarating to be so weightless and free. The colors of the universe were far beyond anything human eyes are able to perceive. Can you imagine a nebula with vision?

While inside Nanci I had forgotten entirely who I had been for eons. I was flabbergasted that I could have forgotten so much of my eternal life. It did not seem the slightest bit plausible that none of these physical lives, and the personality I had become because of them, had been accessible during my life as Nanci. This wealth of memories placed my meager human lifetime into a much larger context.

Former Catholic Doug F's retrieval of memories of other lifetimes lived in the physical world appeared as a series of bubbles like my life review bubble. Doug outlines the process for us: "Then other lives lived started to appear in a visual format. Each life appeared in a bubble; the entire life from beginning to end, in an instant, with awareness of every moment of that life as if it was just lived from birth to death. There were many lives, too many to count. Each life was perfect."[171]

Former Catholic Andy Petro describes a very similar multi-life review. He recalls that "left, right, up, down, [there] are miniature pictures of me, countless movies going on of all of my lives and all of the events of all of my lives. And they are countless. And every time I would focus on one, I would relive it just as I relived it when it occurred on Earth or wherever else I was."[172] Andy remembered that before he incarnated into his current human body he had been on a planet with beings at a higher level of consciousness than we souls are able to have in humans. This coincides with knowledge I received that Earth is considered in the afterlife to be a wild, primitive planet.

Atheist Aaron M also had a flash of memories from other lives, including being a swarm of insects that were eaten by something, and then feeling himself within that predator, which was also eaten by something else. Aaron kept moving up the chain of predators/prey as he was eaten. He remembered living underground as something. His life review displayed his past lives as an endless string of experiences that all contributed to who he is now. He understood during the review that his real, eternal life consists of all that came before this one incarnation as well as this human lifetime.

Unlike the others mentioned, burn victim Anke Evertz was raised to believe in reincarnation. Anke realized her former lifetimes showed both sides of the coin. Sometimes she was rich and sometimes poor. Sometimes her body was fat and sometimes skinny. Once she was a predator and in another life she was a victim. In one life she was an alcoholic and in another a teetotaler. She explains that she had all the experiences that were possible during each of those other lives. Her lifetimes took place in various times in history.

Fundamentalist Ron Kruger's afterlife presentation of other lifetimes was unique and engaging. He met a group of 50 to 100 spirits, each with its own identity but together comprising a single awareness or single force. They hovered in the air in rows formed slightly below Ron's soul position. In the center of the grouping, Ron saw three "oriental women." All the spirits had humanoid faces. But from their shoulders down, their forms blurred to the point that their arms and legs dissolved (the classic Light Being appearance). They were both male and female and had different nationalities. Ron believes these spirits to be his own past lives, with the oriental women being his most recent. Ron describes the merging of identities unique to being Source that allows 50 to 100 lifetimes to constitute one living entity:

> Each of the spirits had lived once, but the truth and experience and wisdom of each lifetime was integral to the entire group. When each soul returned, their lives were absorbed by all, so there [were] no distinctions between thoughts and attitudes within the group. Each of them shared completely every experience and every knowledge of every lifetime into a single conscience. Like spices and other ingredients added to a Mulligan Stew, each added to the mix, but the resulting flavor was one. I was them, and they were me. There were all of my past, and they were my present.[173]

The spirits communicated with Ron telepathically in a way that he instantly understood their thoughts and emotions. Ron started to merge with his other lives but was interrupted by a call to appear before the Council of Love (see chapter 15, "Light Being Council Meetings").

Former Southern Baptist Wayne H was walking when he was hit by a car and drowned in his own blood. He says that in the afterlife he "was shown a long line of experiences in other realms of realities and on other worlds. It was sometime later I realized it was my past 'lives' review of all existences of which I had been part."[174] Wayne explains his multi-life review in his NDERF account: "I did not have a 'life review.' I had a 'past lives in other realms of existence' review." Wayne lived states of existence beyond anything humans can imagine. "There were beings and objects unlike anything I had ever seen or heard of, even in the imaginings of science fiction writers I had read. I was made to know there were an infinite number of realms of existence and all were part of the One, the Source."[175]

Motorcycle accident survivor Emanuele, formerly agnostic, had a multi-life review that encompassed his origin in Source all the way through multiple lifetimes. He says: "I saw events like the death of a daughter that my spirit identified as being from a past life."[176] He also saw scenes from his current human host's future and was advised that he alone was in charge of his destiny.

Former Hindu Mira Sai got a quick flash of memories of her past lives. Religious seeker Yazmine also learned that she had had many other lives, as we all have, because we are all the Universe—the ONE VERSE. She knew she could review those other lifetimes but did not take the opportunity to do so.

About nine months after I returned from death the first time, I had a few episodes when I popped back into that portion of my afterlife experience where I recalled hundreds or thousands of other lives throughout the universe. I have no idea whether these were flashbacks of my first afterlife visit or separate trips back to the afterlife. During these pop-ins, I relived being a fish trailing a boat's wake because I loved swimming in the bubbles. I also relived being some type of big cat, complete with the sensation that I was walking on all fours with my furry belly swaying. The most complete memory was of being a shuttle pilot delivering supplies from a space station to spaceships too large to dock at the station. I happened to see a reflection of my face as this creature in something shiny but not a mirror. My skin was gray-green and slimy-scaly looking. I had three ridges on my head above my forehead and large eyes with orange irises and two sets of eyelids—one set vertically and the other horizontally.

This lifetime as a shuttle pilot is the only one I have noticed interfering with Nanci's life. When I was learning to pilot a small airplane, my unconscious memories of being a shuttle pilot made me instinctively drop the plane down low and come in for a landing very low and close to the numbers on the end of the runway. This is what I did as a shuttle pilot entering spaceship cargo bays! While it makes sense for a space shuttle, it is very dangerous for a small airplane on Earth.

My afterlife memories of other lives proved to me that personality remains intact throughout the entire incarnation phase of eternal life. I am the same personality in a fish or big cat as I am in Nanci. The difference is that less of my personality showed through those other hosts because of their physical limitations.

Once one has remembered in detail living other physical lives, as other humans and especially as other species, it is impossible to continue to believe that one human lifetime has any great significance. Yet it has tremendous importance to both us and our human hosts. Unlike us, this is the only life our host bodies will ever live. We owe a duty to our host bodies to guide and protect them. At the same time, we owe a duty to ourselves as Source to have firsthand human experiences that add to Source's immense database of Knowings.

10

Human Life Isn't Real

WE ALL KNOW WHAT THE INCARNATION STAGE of eternal life is. We are living it now!

The problem is, many believe what we are living now is all that is possible. We souls inside humans unquestionably accept the belief that we are human, and no more. Everything we see, hear, feel, and otherwise experience convinces us this is true. But expanded afterlife experiencers learned how far off this belief is. Without our eyewitness accounts of the breadth and beauty of eternal life, we souls would forever be condemned to accept a narrowly constructed, plodding view of ourselves.

One of the many surprising things my research group members learned in the afterlife is that physical life is intended to be fun. Unfortunately, our bodies do not always get the message. Sometimes they cling to drama and trauma to the point of choking off all enjoyment in life.

Expanded afterlife experiencers received eternal truths that opened their minds to the illusory nature of human life. This knowledge freed them from the shackles of human beliefs. For me, it relieved the pressure of needing to get everything perfect in this lifetime.

Playacting as Humans

Mira Sai describes incarnation as a temporary landing place until we are ready to return to our real home, which is the Oneness of Source. She reports: "I knew without a doubt that life on earth was just a playground of experience, an assignment from God, a mirror projection of the Divine."[177] Mira further explains:

> The universe is God's theater, and souls are like a vast dramatic troupe that has many plays in its repertoire, which they enact at different times ... and different lifetimes ... in different locales. In one lifetime, a soul will be "wearing the costume [having a particular body] and following traditions of that particular life play," performing its part in a certain drama upon a given planet.[178]

Mira analogizes us souls to waves in the ocean, with incarnation equal to rising up for a short while and then merging back into the ocean as the ocean. This reflects the same in-and-out-of-Source concept that other expanded afterlife experiencers have mentioned. Mira sees human life as having taken place only in her mind and declares that her true reality exists outside of human life.

Leonard wants us to know that Source has a great sense of humor. Once he learned his true nature in the afterlife, Leonard laughed about how seriously he had reacted to events in human life. His advice is not to take human life so seriously because it is nothing but a big drama. Kathryn H also calls what happens in human life more of a drama than reality.

Pamela K, in the afterlife because of an extended apnea episode, describes her awakening to the knowledge that human life is not real by using familiar game references:

> Then I began "laughing" when I realized that "I" was not the personality of "Pam" anymore—in fact, I realized that the entirety of my life had just been a game,

> like Monopoly (I guess I just passed Go and collected $200.00!!, and "Pam" was the equivalent of one of the game's playing pieces, like the shoe or the car!) All the troubles and joys, the accomplishments and limitations that made up the story of my life were revealed to be mere dust—an illusion. I realized that in truth, the "I Am" was who I really was and ALWAYS had been and always would be. What exultant bliss![179]

Andy Petro received the same information via communication with a Light Being, namely, that only the Light is real and not incarnated life. The Being explained in detail that we parts of the Light play various roles, like roles in a Broadway play, when we incarnate on Earth and elsewhere. Andy tells host Kirsty Salisbury in his *Let's Talk Near-death* YouTube interview that he was humorously chided in the afterlife for believing that human life was real and taking it so seriously.

Chantal L agrees. She calls her life as Chantal the human a role she played in a play. She was greatly surprised when in the afterlife she ceased to be that role and awakened as the eternal being she truly is.

Author Natalie Sudman provides a very understandable explanation of how human life can ultimately be considered to be unreal:

> From the comfortable perspective of the [afterlife], the physical life could be imagined as the equivalent of our watching a movie. While it's playing, we immerse ourselves in it, accepting the premise, sharing the emotions, and caring how the conflicts conclude. When the movie is over, we snap out into "real" life. Although emotions might linger, we know that the movie was not "real."[180]

I was flabbergasted when I realized in the afterlife that Nanci's life had not been real for me. It certainly was for my body. I was beyond astounded that I had been fooled so completely into believing that Nanci's life was all there was to my existence.

Real "Reality" Is the Afterlife

Aaron M calls his expanded afterlife experience "more real" than anything else he has ever experienced. Emanuele notes that human life seemed like a dream in comparison with the afterlife. Joan LH adds that not only is the afterlife more real than human life, but also that it is our real home. Sue C quips that her level of consciousness and alertness while in the afterlife was so much greater than that of human life that human life is a joke in comparison.

Expanded afterlife experiencers are not the only souls who learn that the afterlife is more real than human life. Some threshold afterlife experiencers who enter the afterlife, even briefly, acknowledge that it is far more vivid and real than human life. For example, Jeff Olsen writes: "Was it heaven? I didn't know, but it made my earthly existence seem like a foggy dream. What I was experiencing was far more real, far more tangible, and far more alive than anything I had ever known."[181] NDEr Kane D agrees: "The experience of death has been the most real and physical experience of my life and the world here felt cold, heavy, and unreal for some time afterwards."[182]

A couple of expanded afterlife experiencers were given part of the blueprint for how the illusion of human life is accomplished.

Aaron M was told by Light Beings that his human persona was "nothing but a bunch of memories and training that made me think I was a human."[183] While in the afterlife, Aaron was given a guided tour of his human body's brain, where he witnessed pulses of electricity moving from the body's brain to its heart. Aaron then saw, and was told by a Light Being, that the human body operated like a simple machine. He learned that humans are no different than other animals, with all being created equally. I too learned that our bodies share most of their traits in common with other earth animals, and that the brain is merely the computer that runs the biomechanics of the body. I was informed that thinking actually takes place in the soul, not in the brain.

Aaron was told that his former body was just a form Energy had taken and that it was time for it to be reabsorbed into the whole.

Aaron fought reabsorption because he wanted to return to his current physical life. At his insistence, Aaron came back to earth after his repeated protestations that he was not ready to return to Source. We are all the wiser for Aaron's determination to return and share his afterlife memories with us.

Where I was in the afterlife seemed far, far more real to me than anything in human life. After I remembered all of my other lifetimes in the physical world, it should have been obvious to me from the evidence that my life as Nanci was not the whole reality of my life. Yet, I did not come naturally to that conclusion. I had to be told mentally that I had separated out part of myself, one of my multiple simultaneous levels of awareness, to incarnate into Nanci while the rest of me stayed in the afterlife. To the part of me still in the afterlife, the incarnated part felt similar to what it feels like in human life for an arm or leg to fall asleep. These Knowings planted into my mind concluded with the fact that human life is not real—except to humans.

I had a difficult time accepting that Nanci's life was not real. Source comforted me with assurance that the experiences Nanci and I had together were absolutely real, as were the emotions we both felt. The life we lived together was real for her and was all that would ever exist for her. But Nanci's lifetime was a mere speck in the vastness of my eternal life. Human life was not my real life in the sense that it was more like a dream, or a phase I went through, rather than the totality of my existence. In other words, human life in context is not the be-all and end-all we think it is.

It simply never occurred to me before I died the first time that if I have an immoral soul, which I did believe, that it could have existed *before* being born in Nanci. I suspected reincarnation was true, but never tried to reconcile that with my belief that I was created to be Nanci. The belief that I am human was the bedrock of my whole life. Finding out in the afterlife that I am not human blew to smithereens everything I knew about life. On the other hand, it gave me a far more comforting perspective than looking at life from the bottom up. Now I see all of eternity equally. I no longer feel like a victim because I know I create my own path in life as part of Source.

I could not escape the feeling that I had been an idiot to have ever believed human life is real—when compared to the far deeper sense of reality that exists in the afterlife. But learning that another part of me had remained in the afterlife—what I consider to be *bilocation*—and was there to help me through human life helped ease the pain. Learning that my loved ones are also bilocated, with one part incarnated and the rest of their Energy and personality in the afterlife, removed the worst sting of suffering through the deaths of my father, mother, brother, stepmother, aunt, and cousins. Even in the depths of grieving I knew that we were still together in the afterlife. The whole of them and the unincarnated part of me lived in unconditional love together until the rest of me returns to the afterlife. This new information helped me see the vast continuum of life beyond Earth's horizon.

The identities and personalities we believe belong to the human body we inhabit are actually our eternal selves–the mental characters/personalities Source created uniquely for us. We remain ourselves, with the same personality and character traits, loves, hates, memories, and emotional baggage, regardless of what type of physical matter we choose as an incarnation host. It is only the physical limitation of our chosen form that limits our ability to show our true selves when we inhabit creatures and things.

I have concluded from my afterlife visits that the single greatest impediment to enlightenment is an unwillingness to give up the belief that we are our bodies. Many spiritual beliefs caught up in this reluctance theorize a spiritual or astral human-like body for discarnate souls because humans cannot imagine life without a body. This kind of thinking may be responsible for our failure to accept that we souls can and do incarnate into any type of physical matter throughout the universe—creatures, plants, inanimate objects, gaseous substances, everything. And it inevitably leads to the erroneous conclusion that we remain human in the afterlife, which restricts the types of blissful eternal lives we can imagine but which extended afterlife experiencers have actually enjoyed.

11

Receiving Knowings

ONE OF THE HALLMARKS OF RETURNING TO THE afterlife, as opposed to just getting out of body,[184] is having unfettered access to the database that makes Source All-Knowing. Many souls that have progressed at least partly into the afterlife mention suddenly knowing everything about everything. Some NDErs even start to receive Knowings downloads while still in the crossing over phase, the stage in which the soul feels itself in transition from the physical world to another state of existence in the afterlife.

"Knowings" is the term some of us use to describe the information obtained directly from Source while in the afterlife. Knowings differ from *knowledge* as we use that term in human life in that Knowings are downloaded whole directly into the soul's mind.

Humans gain knowledge through study or experience, the application of effort to become familiar with a field of study. Receiving Knowings is like having everything that could ever be known about a topic suddenly flood your mind, complete with total understanding of it and the sense of having learned it through personal experience. As Joan LH recalls: "I simply KNEW things without hearing a single spoken word."[185] Eternal truths are planted directly into our minds via Knowings when we are in the afterlife.

How Acquiring Knowings Feels

Expanded afterlife experiencer Anonymous explains that communication in the afterlife leaves no room for mistakes. No misunderstanding is possible. No double meaning implied. No deception. Nothing as primitive as sound waves or words are used. DW agrees, informing us that communication with the Source phenomenon "was not with spoken words but more like with complete thoughts with ... no possibility of misunderstanding or evasions."[186]

Anonymous does not think telepathy was used by the Light Being conveying Knowings. Rather, he says: "I understood what it was saying not with my ears, not with my mind, but with absolutely everything; every molecule in my being knew what was being conveyed."[187] Peter N agrees and adds: "It was clear that the being that was producing the communication wasn't guessing as to the truth of its answer, it was utterly clear that it knew what it was talking about. No ifs, no buts, no maybes, it knew."[188]

Peter N's sensation of receiving Knowings was similar to Anonymous's: "I was being given the essence of an understanding (not just receiving what, in ordinary life, we would call an answer)."[189] Peter describes the communication method used in the afterlife as one of direct transference of thought and feeling. He advises that there are no words actually sufficient to explain it. Peter rejects the common human understanding of thought, explaining, "many people, like me, might automatically regard words as necessary to 'thought.' There this is not the case. Words are not required for thought or communication."[190]

Wayne H acquired Knowings this way: "This was not hearing words in my mind and translating them into thoughts; this was knowing as the other presence knew, an instant sharing of knowledge."[191] In his expanded afterlife experience report, suicide survivor Henry W explains that he received Knowings "like a running narration in a film," answering his questions before he even asked them.[192]

Expanded afterlife experiencer Yazmine S received Knowings too. She confirms the totality of the sensation: "I was experiencing being in the knowing of everything. I was aware of all and everything that is, was, and ever shall be. It was a feeling of great wonder."[193] William W sums it up as simply "I understand and I know."[194]

NDEr Burke (not part of my research group) provides us with a three-dimensional description of what the sensation of receiving Knowings feels like:

> I was immediately bombarded with information that came to me from all directions, through multiple dimensions (as so it seemed).... It was as if my being had become a vacuum, opening up, allowing everything that ever was, or was to ever be, inside.
>
> I now understood everything that ever was, or was ever to be. I didn't need to think or even question anymore. I was complete.[195]

Some NDErs and threshold afterlife experiencers also received Knowings while in the afterlife. Of the 420 near-death experiencers Dr. Jeffrey Long studied for his 2011 book *Evidence of the Afterlife*, 56 percent reported receiving unidentified special knowledge or purpose and 31.5 percent said they seemed to understand everything about the universe. Unlike expanded afterlife experiencers, however, Dr. Long's research subjects often interpreted what they learned via Knowings as affirming their preexisting religious beliefs. Expanded afterlife experiencers recognized new eternal truths.

Topics of Knowings We Received

Everything About the Universe

Most of my subjects submitted their afterlife report to the Near-death Research Foundation's database at nderf.org. The NDERF format allows the submitter to write as much personal narrative as he or she wishes to share. It also asks the submitter to answer a lengthy questionnaire. The NDERF questionnaire asks: "Did you suddenly seem to understand everything?" The submitter has to choose from only three possible answers. They are: 1) no, 2) everything about myself and others, or 3) everything about the universe. All but one of my subjects chose the third answer.

Robyn sums up most of my research subjects' depth and scope of Knowings: "I knew everything that had ever existed in the universe, and everything that ever will exist."[196] In short, we became All-Knowing as parts of Source. Notice that Knowings contain not only currently existing knowledge, but also everything that can be known in the future. Unfortunately, few details about future Knowings are brought back to human life. My research disclosed no NDEr or expanded afterlife experiencer who recalled enough about future information to, for example, create a new patentable product or cure a life-threatening disease.

Miscellaneous Topics

Some members of my research group recall bits and pieces of Knowings on specific topics in addition to having the sense of knowing everything there is to know in the universe.

Andy Petro knew about the creation of universes, solar systems, and planets.

Bridget F reports feeling like her head was thrust into a giant stream of knowledge that allowed her to understand everything about the universe, as well as about cosmology, biology, spiritualism, consciousness, beingness, non-beingness, physics, mathematics, Latin, law, and everything else there is to know.

Aaron M was given Knowings about how and why the universe was made and how it is all connected. Bobby R likewise saw the thread that binds everything together and noticed patterns he had never considered before. DW received Knowings that allowed her to understand "how things were put together, why they are that way, how they had to go and where she fit into the design."[197]

Aaron also learned the meaning of life, that there is no life plan, and that all that matters is that we live and not fear death.

Anonymous received Knowings from the Entity (his name for

Source) to the effect: "Do not worship me!"[198] He was told that Source has already learned all it wants to learn about worship. In Anonymous's opinion, praying is not going to work. He adds that he was told we have free will and choice, life is not set in stone, and that the Entity is not going to intervene in our lives. Anonymous says that this world and what we do with it is all up to us; God is not going to take care of things for us.[199]

Chantal L gives a broad description of the kinds of Knowings she received: "I remember knowing everything. Not so much from an intellectual point of view. I knew what it was like to be a flower, to be an animal, to be an insect. All the knowledge of the universe was inside my being."[200] She explained that she had become part of a collective consciousness (Source) that included billions and billions of others, and that merger with them had provided the knowledge. Chantal clarifies that Knowings are "not intellectual but experiential" because the collective consciousness of Source retains all of the experiences of everything in its memory.[201] Included in that memory are the experiences of flowers, stones, and inanimate objects, which Chantal reveals have consciousness as well.

Demi B's greatest memories of Knowings include that love is all that matters, and that everything and everyone in the universe is the same, which is love. She was told she is exactly the same as every blade of grass. This Knowing confounded her and propelled her into studying science, physics, biology, chemistry, and math in college in a quest to find the answer to how that could possibly be true. Years later, she realized that the sameness is because everything that exists is energy at its core.

Doug F learned that everything is already known and there is nothing for us to learn. Therefore, life on earth is not a school but rather an experience. He learned there is no heaven or hell as religions preach.

Henry W's Knowings came in the form of voiceless words plus experiences. He lists specific areas of knowledge he received, including understanding the formula $E=MC^2$, learning Earth is like a Petri dish designed to raise humans, the fact that God cannot be proven by scientific means, an explanation for why bad things happen to good people, and whether ghosts are real.

Clinically dead four times from cardiac arrest, blood loss, and having a stroke, Jennifer J recalls her own specific Knowings: "I was aware of suddenly having infinite knowledge. I knew all languages, ALL languages, and all religious thought, all everything."[202]

Leonard remembers getting answers to all the questions he had ever wondered about, but he was unable to bring those answers back with him. He does remember Knowings about who created the universe and how. He recalls learning that we are eternal beings that have always existed and always will. Other Knowings allowed him to remember how the cosmos works, physics, and everything else about the universe. He clarifies: "Oh yes, I did not learn it, I remembered it!"[203]

Malinda K adds even more depth to how Knowings operate in the afterlife. She had a very intimate knowledge of all things, both seen and unseen, because she actually *was* them. Chantal L and Robyn felt the same way—that they actually *were* the knowledge they received.

Mira Sai reports that universal laws rapidly unfolded and flowed into her.

Natalie Sudman's book touches on some of the many concepts she learned about in the afterlife, including a vast amount of information on manifesting physical reality, that the belief structures of karmic law and religious punishment have no place in her experience, and that the afterlife is a state of mental focus rather than a physical place.

Robyn's Knowings changed her beliefs about science. She now calls science a mere activity of the mind that cannot come near to understanding the totality of the universe.

Ron Kruger, on his way to meet the council of Light Beings monitoring his mission, received Knowings on a wide variety of subjects, such as that the ultimate achievement of science is to prove the existence of God, human love is limited, and everything in the afterlife is infinite. His Knowings on human use of language is particularly instructive. He learned that human spoken language is primitive, unreliable, and used more to deceive ourselves and others than for communicating the truth. Humans use words to label, distinguish, and separate everything, resulting in both a loss of true understanding and loneliness.

Valeska was given the meaning of existence and received many "flashes" of information. She concluded from what she learned that everything is simply love.

Yazmine knew everything about the world, animals, insects, ocean life, plants and trees, water, air, fire, the wind. She learned that the sky is alive. Yazmine describes Knowings as a feeling beyond normal knowing, perhaps better described as boundless wisdom.

Shortly after I first entered the Light, Knowings started flooding into my mind at a firehose pace. The largest block of knowledge I can now recall was about human traits and how they differ from our true spiritual nature. The second largest block is that manifesting is the divine power to create physical reality. Among the hundreds of other pieces of information I received, I remember bits and pieces of Knowings about scientists, linear time, math as a human language, mental illness, that humans are capable of living without a soul, no human is one hundred percent male or female, and Source could not care less about human sexuality or what religions call sin. I remember more about suicide, reincarnation, what humans mistake as love, the death and crossing-over process, psychics and channels, and a wealth of other topics.[204]

After my multi-life review, I realized that every category of knowledge, plus past, present, and future facts about anything was available to me via downloads of Knowings. I eventually discovered that I could focus my attention and intention on particular questions and access Source's answers to them while I was in the afterlife. I asked: What is God? What am I? What is the purpose of life? What does God expect of me? Where is heaven? Where is hell? And what is the one true religion (Catholics think their religion is the only true one)?[205] Brief summaries of the answers I received follow, with more detail in chapters of this and my previous books.

What is God?

Before I died, I believed the dogma contained in the Roman Catholic catechism. In the afterlife, I received detailed eternal truths about Source and its nature. I understand that Source is the only divine entity. There are no other gods or spiritual beings or angels or religious figures in existence. Source is what humans call God, Yahweh, Allah, the Creator, the Supreme Being, and many other names. Source has no physical matter, looks nothing like a being, and does not act like a human father. It is the Creator of the universe and all that is in it. Chapters 1 to 3 in this book detail not only my own perceptions of Source but also those nearly identical ones offered to us by other expanded afterlife experiencers.

What am I?

Before I died, I believed I was a human being who possessed a soul. A huge download of Knowings about what human animals are, and are not, and what I actually am proved otherwise. I learned that I am the entity we call a soul, not the human animal I inhabit and left behind at death. Chapter 5 describes in detail how we souls were created and why.

I was still me in the afterlife. I still had the same personality, character traits, loves, hates, and emotions with which I was familiar while inside Nanci. The only aspects of my former life that were missing, besides the obvious physical body, were pain and human survival and fear-based emotions. Several others in my research pool also specifically remarked in their narratives that they were still themselves in the afterlife, just as they knew themselves on earth. These include Andy Petro, the Anonymous Experiencer, Joan LH, and Robyn.

I was told that I am on Earth because I chose to be. I learned through Knowings that Earth is considered by Light Beings to be a primitive planet with wild and wooly creatures roaming it. It is viewed a little like how we now think of the Wild West in the early days of colonizing the United States. Souls that choose to incarnate on Earth do so because they are looking for a greater challenge than possible in other incarnations. They want to test themselves to see whether they are able to show unconditional love in the most dire and difficult of circumstances. This is why NDErs are so often told that "love is all that matters" as a goal. In the words of NDEr Dannion Brinkley: "Those who go to earth are heroes and heroines, because you are doing something that no other spiritual beings have the courage to do."[206]

At the end of my first expanded afterlife visit I awakened to the knowledge that I am Source itself. Chapter 3 reveals how others like me learned they too are Source.

Chapter 4 details the Knowings we expanded afterlife experiencers received about our creation.

What is the purpose of life?

Chapter 6 is devoted to this topic.

What does God/Source expect of me?

Nothing. Just to experience whatever there is to experience in the physical world. We mental character aspects of Source have the same full range of free will as Source does because we are Source. We choose where and how and in what host we want to live and for how long. We choose when to stop living as though we are separate identities and when to reawaken as the core entity Source.

Where is Heaven?

I thought at the time I asked this question that I was in heaven, but I wanted to be sure. Knowings informed me that heaven is a state of existence—a state of mind, not a physical place as religions portray it. Heaven has no physical matter to it. It is not an environment but rather a state of awareness and beingness.

Light Beings in "heaven" appear to NDErs as luminous bodies of Energy. Or they may take on the façade of an angel, human loved one, or religious figure. Their reason for using familiar human religious icons is to serve the needs of the newly arrived souls—to make them feel as welcome, accepted, and peaceful as possible after their traumatic exit from incarnation. Knowings explained that the earth-like manifestations NDErs perceive in heaven serve this purpose as well. This is why the scenery is different in each report but follows the same pattern—a setting a human would find to be heavenly or comforting.

Where is Hell?

I could tell from Source's memories of Creation that Source was alone before it started creating the universe. There was no counterpart spiritual entity that humans call the Devil or Satan. Moreover, while I watched and remembered Creation, it was obvious that Source did not create a Devil or Hell. There would be no purpose for it. Why would Source want to punish a soul for doing exactly what it was created to do? Knowings informed me that evil comes from unrestrained humans merely acting like the animals they are, not from the influence of evil spirits. Emanuele and Doug F also learned that there is no hell. Pastoral counselor Joan LH adds: "There is no 'heaven' and no 'hell' – except what we do to ourselves. I suspect the judgments we hold against ourselves is the real 'hell.'"[207]

What is the One True Religion?

Knowings that several expanded afterlife experiencers received answered this question unequivocally: There is no one true religion. Religions each have kernels of truth in them but consist mainly of human-created myths, superstition, speculation, and fantasy centered around an anthropomorphized deity that supposedly acts very human.[208] NDEr Jean R (not part of my research group), another Catholic, also asked what the right religion is while she was in the afterlife. The answer she got was:

> But religion in and of itself is destined to always be

> interpreted by man. Their interpretations are often lacking real clarity and have many falsities in place too. One of the things I was told was to look always to "who benefits?" in rules made by religion. If it is particular men, or the power structure itself, chances are that it is not really something of God. Many rules are definitely man made and put there to benefit either the structure or those in charge of the structure.[209]

Expanded afterlife experiencer Jennifer J adds: "I encountered information that human interpretation of biblical or religious nature is often wrong, made by man and described or explained for man in human terms, but mostly wrong."[210] Anonymous received Knowings that while the core tenet of religions, such as do no harm, is valuable, religion's messages generally have been tainted by humans who like to control other people.[211] He adds that Jesus and angels are not going to save you and that there will be no rapture.[212]

What Joan LH, who was a trained Catholic pastoral counselor at death, remembers most is that religions have it wrong and make it all so complicated when it is really very easy. She explains:

> It's interesting to me that what I most clearly remember is that "religions have it all wrong: all religions." It was an immediate Knowing that we've been focusing on rules, on suffering, on trying to "get right with God" when we've always been "right" and in the heart of God. It wasn't just Christianity that I felt had it "wrong" but all religions.[213]

Internationally acclaimed NDE researcher Dr. Bruce Greyson mentions several times in his book *After* that some NDErs he interviewed were confused by the conflict they saw between what they personally experienced in heaven and what they were taught by their religions to expect after death.

Most expanded afterlife experiencers abandoned their former religion after having their experience.

Eternal Life Has Multiple Stages

I received Knowings that the state we call heaven is one of many states or phases of an eternal life far greater than any human could ever imagine. Eternal life has stages or phases just like human life does. Humans live as an infant, toddler, child, teenager, adult, and senior. The eternal life phases I remember learning about are:

- the incarnation phase we are all experiencing now;
- the "crossing-over" phase, which can include moving through a dark area and/or a tunnel-like area;
- going into the Light or other demarcation of a higher state of existence;
- the transition back to spiritual life phase, which includes an in-depth life review;
- the living vicariously via merger into other Light Beings phase;
- the awakening as Source phase; and
- the optional missionary on Earth phase.

Each phase has multiple parts. I am aware that many other phases or stages exist, but I do not recall them.

Many NDErs who meet the scientific definition of the term near-death experiencer get out of body but do not begin the process of crossing over. They are literally near death but not post-death. They are still in the incarnation stage. Their accounts are thrilling and life-affirming. They give us hope and reassurance that life does not end when the body dies.

Some souls not only get out of body, but they also begin crossing over back to our home, what most call heaven. The oral and written reports of these courageous souls form the core database of NDE research. Without their testimony we would have nothing but speculation about what awaits us on the other side. Their willingness to share their most intimate moments after death gifts us all with knowledge and comfort to literally last a lifetime. Yet, most of the

NDErs who begin the afterlife experience witness only parts of it before they are halted in their progression because of resuscitation or other reasons.

Expanded afterlife experiencers by definition complete the transition stage and reach later eternal life phases before returning to their former bodies. This book compiles their amazing journeys.

The sensation of being All-Knowing we have deep in the afterlife is exhilarating. Unfortunately, we souls cannot retain all that data upon returning to earthly bodies. Often, it is all we can do to simply recall the topics of information we received. This is particularly true if the expanded afterlife experiencer has also been gifted with all the data stored from Earth's past and future.

12

Earth's Past and Future

ANOTHER ELEMENT OF EXPANDED AFTERLIFE experiences is being shown Earth's or humans' past history and potential future events. A few threshold afterlife experiencers got this vision as well.

Visions of Earth's Past

The NDERF-created questionnaire that most of my research subjects responded to did not include a question about whether the experiencer had seen anything of Earth's past. Very few NDErs or threshold afterlife experiencers included visions of Earth's past within their written narrative or video account online. Consequently, the prevalence of this activity is unknown.

Expanded afterlife experiencer Aaron M was told about history and the future all at once, in the mental voices of thousands of Light Beings. Bobby R, who was four years old and blind when he died, saw a thousand years of other people's lives while his life review was playing out in the afterlife.

Upon receiving all of the Knowings answering my seven most important existential questions (see chapter 11),[214] I became angry that this information had been withheld from me during my lifetime within Nanci. I assumed that everyone else on Earth knew these eternal truths. I was insulted, thinking that my parents, priests, and nuns did not judge me smart enough, or sophisticated enough, or maybe even worth enough to be told the truth. I felt as though my religious leaders and parents had lied to me because they thought I could not handle the truth. I now believe the reason I was not given the true answers to my seven greatest questions while incarnated into Nanci is that only a few other afterlife experiencers on Earth knew them. I had never encountered an afterlife experiencer before my death in 1994.

In seeming response to my anger, Source showed me a vision that was like a documentary movie of Earth's creation and its future up to the planet's eventual demise.[215] The vision was accompanied by Knowings explaining what I saw as I viewed it, and something like TV news chyrons running across the top of my visual range that added labels for time periods.

I watched gases and rocks coalesce into the planet Earth, the chemical reactions that gave rise to organic matter, and the origins of human animals, as well as every other life form on Earth. I witnessed the mass infusion of Light Being souls into humans and the subsequent birth of religions in mankind. Religions and spiritual beliefs changed dramatically as humans themselves developed more sophisticated societies.

During my world history review, I saw that Source has given us souls incarnated on Earth millions of messengers of the truths I had just learned through Knowings. None of them were/are religious leaders, kings, pharaohs, gurus, prophets, or other self-appointed experts on gods and what they demand. Some of them, however, are famous for other reasons, such as being popular musicians, singers, poets, artists, and TV show producers. I recall seeing that Gene Roddenberry, who created the *Star Trek* TV series, was one of Source's messengers of truth. His Starship Enterprise characters displayed many Light Being traits, including a lack of interest in fame or fortune; a respect for the equality among the sexes, races, and species; and dedication to doing what is best for the most people involved rather than oneself. Many other messengers are the same as you and me, regular people living ordinary lives but sharing through their words or deeds the truth about Source, life, death, and eternal life. Any threshold afterlife experiencer who received pure Knowings without human interpretation bias is a messenger from Source. And there are millions of us.

In a similar fashion, Ron Kruger was shown what he calls a newsreel movie of Earth's past events while he was present in his council meeting.

Visions of Earth's Future

Earth's future as it appeared in 1994 was also shown to me in the documentary vision, but I remember little of it. I paid less attention to the future than the past because I knew I would not be going back to human life—or so I thought. The memories I was able to recapture as I whirled back into Nanci's body are reproduced here from my book *BACKWARDS: Returning to Our Source for Answers*:

- I remember seeing an aerial view of India with the mouth of the River Ganges darkened, as if it had been flooded or sank out of sight. My guess upon resuming human life was that the land mass would be destroyed. But I was wrong. I saw the same aerial view in the newspaper on January 26, 2001. It was a satellite photo of 32 million Hindus crowding the confluence of the Ganges and Yamura Rivers for the millennium celebration of the Kumbh festival. The presence of so many people darkened the Earth in the view from space, making it appear as though the land had submerged.

- While simply observing planet Earth, I noticed a tenth planet outside Jupiter orbiting our sun. It appeared to be mostly rock and ice. I found it amazing that human scientists could miss an entire planet in our solar system. Michael Brown of the California Institute of Technology discovered the planet, informally named UB313 at the time, as reported in the newspapers in March 2004. The dwarf planet is now called Sedna. I saw many other planet-like celestial bodies beyond the solar system I learned in elementary school, many of which have since been discovered.

- Habitation on Earth spans three time and life cycles identified to me as "epochs," each lasting millions or thousands of years. We are currently near the end of the Second Epoch. My best estimation in 2007, when my first book was published, is that the Second Epoch will terminate around 2013 to 2015 when a more pronounced transition to the Third Epoch will begin. I cannot be certain of the exact date because such details did not survive my reentry into the body, despite my best efforts to retain them. But my sense is that the current Epoch ends during my remaining human lifetime, which I estimate to end around 2033 (based on a second life review I had while back in the body).

- The transition to the Third Epoch will not be as abrupt or devastating as that heralding the Second Epoch. Entire species, including humans, will not be extinguished, although the human population will be decimated by natural disasters and disease.

- I saw no global war destroying mankind. No nuclear wars at all.

- Hundreds of disasters played out on my spiritual mind's viewing screen, starting after the turn of the millennium and becoming more frequent and more intense as the Second Epoch comes to a close. Included were earthquakes, tidal waves (tsunamis), floods along the coastlines and riverbeds, freakish weather changes, global wildfires, glacial melts, volcanic eruptions, and the spread of new epidemics of old bacteria and viruses that had mutated. I recognized one such disaster in 2002, an earthquake in Japan, looking at pictures in the newspaper. The 2005 tsunami is another example of future memory I retained that came true. Many more will follow. The mutations of cold viruses into SARS (including COVID-19), and Avian Flu viruses into deadly strains are consistent with what I saw of the transition's diseases. So is global warming.

- I saw the United States in relief-map format with parts of the East Coast missing or flooded, including New York City and bedroom communities up and down the East Coast sunk below water level; Florida's peninsula was missing; the Mississippi River overflowed such that Louisiana was under water, along with half of each state to the immediate north; and the West Coast was missing from just south of Big Sur to Mexico. Mountains formed the coastline of Southern California, but I do not recall which mountain range.

- A global aerial view disclosed that Japan was gone. Simply gone. I do not recall the nature of the responsible cataclysm.

- I saw the collapse of the world's financial system with the U.S. federal government going bankrupt. I surmised the cause to be government bail outs of various industries hit hard by natural disasters.

- The upcoming transition to the Third Epoch does not constitute the End of Days, as that term is used in Christianity. A Third and final epoch of peace and spiritual enlightenment on Earth follows the transition from the Second Epoch. The human population is greatly reduced, inasmuch as only those emotionally evolved enough to participate in this tranquil era will survive the transition. Those who did survive appear to be of two types: those with sufficiently mutated DNA to be called a new human species, and those who accelerated their own personal growth.

- I saw that a third human species was currently being created via mutation of DNA of the second species, the humans that most of us presently inhabit. The difference in DNA is slight. I saw no way to distinguish this third species from the current version.

Ron Kruger was shown a similar future vision that he describes as a sort of newsreel of chronological events. He writes that: "We are presently undergoing the Transition, the labor pains if you will, of the birth of the Age of Benevolence."[216] This sounds very much like what I experienced as the coming of the Third Epoch.

Aaron M and Andy Petro were shown visions of Earth's future, along with the history and future of the entire universe. Aaron saw "places that haven't been made yet, but somehow they still existed at the same time," leading him to believe the future is no different than the present because time does not exist.[217] Aaron saw constant creation and demise at the universal level, perceiving new universes forming and ending constantly.

Bobby R witnessed humanity's future and his part in what the species becomes. He reports that humans do not become extinct or kill each other off but says he will take to the grave his memories of what does happen to humankind.

William Horden was shown scenes from Earth's future presented as alternatives or forks in the road that he calls "emotional choices" that need to be made. He clarifies that these are not predestined events.

Yazmine S also had awareness of the future. She saw that people want to wake up, and that incarnated souls with high awareness are trying to start a chain reaction, but others in high places do everything they can to stop it. She warns that many earthquakes, rising oceans, and 600 mph winds are on the way.

Bridget F, Dea M, Henry W, and Leonard all remember having been shown scenes from Earth's future but cannot remember the events themselves. Pamela K responds to the NDERF questionnaire that she saw scenes from Earth's future, but she does not describe them.

Personal Future Visions

A few expanded afterlife experiencers previewed their own futures while in the afterlife.

DW saw scenes from her own human future. After she returned to her body, she would have day dreams, visions, or night dreams that were particularly vivid. Later, they would happen in her physical life.

Gwen J says she traveled into the future to see what her life would be like there to give her more information to help her decide whether to return to her body or stay in the afterlife.

Emanuele saw scenes from his own future family. He notes, however, that those scenes may not play out because he alone is in charge of his destiny.

As part of her life review, Sue C saw scenes of her future life as an older woman as well as scenes that spanned her entire potential human lifetime. Similarly, Bobby R watched a life review consisting of 128 years of human life even though he died at age four. It included previews of his mother's and sister's future human lives, scenes that greatly disturbed him for the rest of his life after he returned to this body.

I have sporadic memories of Nanci's future shown to me during my second life review, including seeing a second marriage that failed and my older sister's death. I decided then and there to change my life and did not remarry.

Afterlife experiencers have learned through experience as a death survivor that the futures they witnessed in the afterlife may or may not come true. Knowings informed me that the reason for this is that manifesting is a moment-by-moment process always subject to free will. We can change the future by simply manifesting a different present.

13

Living Other Souls' Lives Vicariously

ONLY WHILE DEEP INSIDE THE AFTERLIFE CAN WE souls experience a life phenomenon that has no corollary within physical life. There is no terminology to adequately describe it, or even label it, because it is so foreign to everything human. I call it living other souls' lives vicariously for lack of something more accurate. The other expanded afterlife experiencers who have lived this phase of life have also done their best to use language common to physical life to hint at the beauty and mind-blowing exotic nature of literally sharing another's lifetime memories as though they were lived firsthand.

Henry W felt his heart slow down until it stopped during a suicide attempt. While in the deepness of the afterlife, Henry sensed unknown others around him that he soon began to perceive as a chain of individual, small golden orbs of Light all linked together in what he calls God. Henry could hear these orbs of Light communicating telepathically about their perceptions and sensations while living on Earth. Henry describes for us what I recognize as vicarious living:

> They shared not only in words, but in sharing the experience. If one orb couldn't understand, it disappeared and then reappeared. The orb somehow went back to Earth and experienced that "life" to further understand.

> ... These orbs or rather "souls" would leave this realm, detach themselves with this universe, and return to the universe of our Earth. There they would live and die, then return and share the experience with all the other souls. A soul that could not understand the experience could go and live that life also to experience that life. I learned we have many lives, past, present and future.
>
> These souls, our souls cannot experience certain things like pain, sorrow, hatred, and anger. Though these are negative things, it was important for them to understand and experience them.[218]

I can truly see why Henry describes vicarious living as returning to the physical planet Earth to repeat another soul's lifetime. That is what it would look like to an outside soul observer. But having repeatedly been through this process myself from the inside, I understand that actual incarnation is not involved. There is no reanimation of a dead human body. This is not a walk-in soul situation. No new soul is stepping into a living person. Nor is this possession. No human is involved in the process at all—just the soul's memories of the experiences that a human had lived.

Vicarious living feels like merging souls together, combining all thoughts, memories, sensations, and emotions. Combining minds and individualities. Literally absorbing another into oneself and vice versa. In my experience, vicarious living is a mental merging more sophisticated than but slightly reminiscent of the Vulcan mind meld of *Star Trek* TV fame. Merging spiritual Energy is so intimate and compelling that it constitutes a sublime bliss all its own. Vicarious living is still as real as literal incarnation, however, but without the time commitment involved. As Henry explained, an orb would disappear, live a human lifetime, and reappear in the afterlife while he was watching.

Ron Kruger encountered a group of 50 to 100 spirits in the afterlife. He recognized them and knew that each one had lived once and returned to the afterlife to have their experiences absorbed by their whole group, a form of vicarious living. Ron reports: "Each of them shared completely every experience and every knowledge of every lifetime into a single conscience. Like spices and other ingredients added to a Mulligan Stew, each added to the mix, but the resulting flavor was one."[219] That *one* equaled Ron himself. Ron explains that these 50 to 100 spirits were him, he was them, and all were one. Ron believed they were his past lives in spirit form. Thus, Ron experienced instant vicarious re-living of his own past lives as though they were the experiences of other souls separate from him.

Like me, expanded afterlife experiencer Peter N personally lived this stage of afterlife, instead of merely watching. Peter crashed into a parked car while he and a friend were riding a motorbike. The impact was violent enough to eject the friend over the top right side of the car and to whipsaw Peter back and forth over the handlebars until he flew backwards onto the road. He died from shock just as nurses arrived on scene and tried to help him. Peter calls Source's mental characters, what I call Light Beings and Henry calls golden orbs, "stars." He saw them all around himself in the afterlife. Peter describes how these stars merged into his own Energy: "I was being permeated by the very thought-feelings of these beings so that I could quite literally feel them inside me (though I knew them to be all around me.) … It needs to be understood that I really am talking about the mind of 'something else' entering one."[220] Peter characterizes the experience as being part of a meta-mind, to which he was being reintroduced upon his return from human life. Peter says he cannot bring himself to call these feelings and thoughts remotely human-like.

William Horden died at age 53 of a massive heart attack that he considers to have been a genetic timebomb exploding in his chest. William describes Source's collective nature as consisting of multiple spheres of aware light comprising one large sphere of aware light. He recalls merging with other souls as "whenever another sphere of aware light came into contact with me, there was an immediate and spontaneous exchange between us of our respective Memory and Understanding." He adds: "when we come into contact there [within Source], all that we know and all that we are passes uninhibited between us in a natural and open communion of shared being. Spheres of aware light touch and so exchange the totality of their experience and assimilate one another's experience into their own."[221]

Once I had calmed down and digested the vision of Earth history and future I had been shown (see chapter 12), I was again met by my five eternal Light Being friends. They somewhat pulled me with them into the vicarious living phase of eternal life. I spent quite a bit of "time" in this activity, though *time* is clearly not the right word. Afterlife experiencers unanimously agree that time does not exist in eternal life.

My Light Being friends mentally told me I had to become proficient at merging my Energy into theirs before I could proceed to the next phase of eternal life. They communicated that I needed this practice in order to be able to withstand entering the core of Source. My friends explained that Source would feel like a Collective Being to me because of the multitude of Light Being characters in its mind. They warned me that the sensations of awakening to awareness as Source would be a shock to my system if I were not prepared for it. I had no idea what they were talking about, but I liked the idea of living their respective physical lives vicariously without having to go through the pain and suffering associated with incarnating.

I experienced this vicarious living phenomenon from both the outside and the inside.

The outside perspective instance was during my expanded life review. I was at the center of the bubble of Nanci's life events, which were swirling around on the outside of the bubble, when I realized that my Light Being friends were popping into and out of individual events in Nanci's lifetime. Each one would blend its Energy into a scene and live the events taking place in that scene. Sometimes the Light Being would participate in the scene as though it were me, viewing it and acting within it from inside Nanci's body, just as I had. At other times, my Light Being friend would be enjoying the scene as itself though still from inside Nanci's body. It amazed me. It also freaked me out. I wish I had paid more attention to what it felt like to have other souls popping into and out of my life like that. I was more interested in reliving my own past and other physical lives, which I could see and remember at the same time.

Much later in my afterlife visit, I experienced the vicarious living phenomenon from the inside while with my Light Being friends.

The task itself was simple. Just focus on merging into another Light Being and it occurs, with the other's permission. But the feelings and sensations associated with combining Energy essences would terrify a human. The first time I tried it I was able to see, as though through memory, all of the physical and non-physical living my Light Being friend had experienced. I could watch any part of its eternal lifetime just by focusing on it. Later, I discovered that I too could pop into scenes from my friend's memories. This felt very similar to incarnating into one of my friend's physical lives, with a twist. I could participate in the scene either as myself, or as my friend. And I could switch back and forth between the two frames of reference.

During this phase, I was able to merge my Energy into the Energy of each of my five Light Being friends one at a time or more than one at a time. Eventually, I could merge into all five of them at once. The aspects of their lives that I lived vicariously were as completely real to me as Nanci's life events had been.

Joining essences with my five Light Being friends was breathtakingly emotional. There is nothing even remotely similar to it in human life. Nothing a human can imagine could provide even a whisper of what it actually feels like. It was the most alien aspect of my afterlife journey. But it was far more real than any empathic or psychic experience I ever had as Nanci. More important, it completely convinced me that I am far more than merely human.

Vicariously living another soul's physical lifetimes appeared to me to be an alternative to incarnating into the physical world. In my experience, it was an optional phase of eternal life that is available after a soul has completed the incarnation phase. One can simply choose to stop incarnating and then choose to enjoy others' lifetimes using a method that is far quicker than entering and leaving the physical universe. This is part of the glory of Source.

14

On a Mission

A visit to the afterlife, regardless of how brief, is overwhelming in its impact. It is life-changing in so many different ways. Often, souls who have been on this journey return with a burning desire to share their experience, feelings, and conclusions with the world at large. According to YouTube data, at one time there were 42 NDE channels on YouTube and approximately 42,000 videos. New channels and videos are added every day.

While they are still in the afterlife, some courageous souls are specifically given a mission to share their experience. Thus, an NDEr with a mission is not a novel occurrence. Two expanded afterlife experience accounts in my research data revealed that those souls had been on a mission in this human lifetime, *before* the body died and went back into the afterlife. So was I. This was an unexpected finding in my research.

Previous Mission

Chantal L writes how knowledge that she had been on a mission in her human lifetime affected her during her afterlife visit:

> I then remembered asking to be born. The moment this memory came to mind, I also remembered having agreed to accomplish a mission. I came to earth to

> accomplish a mission. All of a sudden, I realized that I had not accomplished this mission. My joy and serenity turned into anguish; I had failed to accomplish this mission. I never felt so sad. Sad is not the word, I felt anguish. I felt like a failure. Nothing else I had accomplished while in this life mattered to me, I was a failure.[222]

To her great consternation, Chantal cannot remember what the continuing mission is now that she is back in her body.

Author Natalie Sudman had a meeting with Light Beings monitoring her mission after she was blown up by an IED in Iraq and died. The primary purpose of the meeting was for Natalie to upload to those monitoring her mission all of the data on human life she had collected to date. Her book explains:

> In this new environment, I stood on an oval dais ... addressing thousands of white-robed beings or personalities.
>
> ... Rather than being a classic life-flashing-before-eyes scene, this download [from her to the Light Beings] was a collection that emphasized what might be very broadly understood as cultural and political information. I was aware that I deliberately offered the condensed data in fulfillment of a request that had been made by this Gathering of personalities prior to my taking on this body for this physical lifetime.[223]

Natalie was given new tasks to perform as part of her continuing mission once back in the body.

When I first reunited with my five Light Being friends in the afterlife, they mentally sent me emotions expressing welcome home and unconditional love. And laughter. They reminded me that I had volunteered to come into Nanci in order to perform a specific mission that ended with my death. I never knew what the mission was, but they did. I was to act as a catalyst in other people's lives to help them reach their chosen spiritual goals. My dear friends joked that it was ludicrous for me to have volunteered for this mission because I "made a terrible human." They were extremely curious to know how well I had performed and, so, participated in my life review with me.

New Mission for This Lifetime

Many expanded afterlife experiencers began a mission for the first time, or were given a replacement mission, after their life-altering experience. For example, Bobby R saw in his life review that he would be an author and would help teach children. He has accepted that as his mission for the remainder of this incarnation.

Jennifer J relates that she was told her purpose back on Earth would be to circulate love and what she had learned in the afterlife to as many souls as she possibly could. Her report tells us that the purpose of life is love and the purpose of love is life. She explains that life comes from love. Love comes first and is the breath of life. Robyn was also given the mission "to expose this [expanded afterlife experience] to as many people who are interested during my incarnation here."[224] Robyn's further purpose is to usher in a new paradigm for this beautiful planet, to help save humanity. She is also tasked with getting to know herself, seeking bliss, and learning to be a creator of her human experience by focusing her energy on what excites her.

Joan LH, a former Catholic Pastoral Counselor, has left the church and religion behind and helps people become their authentic selves through love. Part of her mission is also to tell people about her afterlife experience.

Kathryn H's new purpose in life is to work to make earth a better, more compassionate place and to find the heartbeat of life and join with it. She will also celebrate the Knowings she received that we are all One.

Mira Sai, born Arti Gupta, was told to "do the work" when she got back into her body, without anything being specified. She currently lives as a mystic.

Ron Kruger was shown a newsreel-like vision of likely future Earth events. He was told that he could make an impact on, or even affect the outcome of, these events if he returned to his body. Ron explains that each soul is unique in its contributions to life and how it evolves. Unfortunately, Ron returned to his former body with no knowledge of what he was supposed to do to accomplish his mission.

William Horden's previous spiritual beliefs deepened as a result of his afterlife experience, which he interprets within the context of those beliefs. The experience led him to retire from counseling and begin writing as an outlet for expressing the ultimate reality he lived while in the afterlife. His books reveal the human nature of many of his beliefs.

At the end of my first deep afterlife experience, I was given, or assumed, or decided upon (I have never been clear on this point) a new, second mission for myself. It was twofold: to tell anyone who would listen what I experienced and learned in the afterlife, and, to experience unconditional love within Nanci's life. I am fulfilling both parts of my new mission by writing books, speaking, and coaching people on how to improve their manifesting skills.

My understanding from afterlife Knowings is that most incarnated souls do not have a mission. However, any NDEr or afterlife experiencer can be an afterlife missionary upon return to their body. Their common denominator is unconditional love for other incarnated souls who could benefit from an infusion of eternal truths. The experiencers' collective goal is to improve the conditions of human life and the earth habitat by enlightening incarnated souls about their true nature.

15

Light Being Council Meetings

SOMETIMES A NEAR-DEATH EXPERIENCER REPORTS feeling judged by some type of tribunal during their life review in the afterlife. I do not doubt these reports in any way. However, expanded afterlife experiencers unanimously agree that there are no judgments made by Source, any Light Being, or anyone else in the afterlife. This is one of the many differences between the two types of afterlife visits. Leaving the concept of judgment aside, some afterlife experiencers report appearing before a group of Light Beings or other entities during their visits. Such gatherings have an established purpose.

I personally experienced that the activities of those souls with a specific mission to fulfill in human life are monitored by what I call a "council." Souls with incarnation goals, but no particular assigned mission, are not monitored by anyone. The reason for the monitoring is because the amnesia of incarnation makes the soul forget it has a mission, and therefore it may fail to take the necessary actions. Occasional meetings with the soul's council will remind the soul of its mission and bring it up to date on its progress. However, some souls interpret the meeting as a tribunal passing judgment. It can feel like that to a soul expecting human traits and thinking. To some extent, the council is assessing and commenting on the soul's performance as a missionary. I know I felt a bit judged both times I was before my councils. But I can personally testify that there are no consequences if a soul fails in its mission.

I had two afterlife experiences during which I met with a council of Light Being advisors who were monitoring the progress of the mission I accepted at the end of my 1994 expanded afterlife visit, which was to tell anyone who would listen what I had learned and experienced in the afterlife.

Many death survivors make dramatic changes in their lifestyles. I did not—at least not immediately. Consequently, the council monitoring my mission called me back to the afterlife within months after my initial expanded afterlife visit in 1994. Death came so swiftly that I do not know how it happened. I know only that I seemed to be alive as Nanci one moment and instantly transported to the council meeting the next second. Several brilliant Light Beings that were unknown to me seemed to be gathered in a type of formal semi-circular conclave, similar to a legal hearing or tribunal, with my new mission on earth the focus of attention. It was evident to me that each person in the group had a keen interest in how well I was performing, and that they were not even slightly satisfied. My impression is that the council basically told me to "get with the program"—my own crude phraseology for working on my mission. The encounter felt like an intellectual slap upside my head to wake me up, though I am certain they did not mean it that way. I felt that way because I realized I had failed. The council members were there to guide and support me in meeting my goal. I do not remember exactly what they expected me to do, but my conscious mind later interpreted my instructions to be to start pursuing my new spiritual mission.

I backslid into mostly ignoring my mission for another seven years. It took me that long to process and come to grips with the conflict between my Catholic religious indoctrination and what I personally experienced in the afterlife.

A second meeting with council in 2001 was more dramatic and traumatic, probably because I knew Nanci was dying before it happened. Earlier in the day I had been in a hospital emergency department with a splitting headache. I was erroneously diagnosed as having a migraine, treated, and sent home. Later that evening, home in bed, I experienced extreme nausea, chills, and what I thought were convulsions for hours. I once again left my body in the bed and drifted into the afterlife to what appeared to be a hastily called gathering in the Light.

When I died this time there was no dark void, tunnel, pinpoint of Light or other introduction like the first time I visited the afterlife. Instantly, I seemed to be in a conference room with barely defined see-through walls and a fairly solid-looking long conference table. The luminescent Light Being council members were still coming into the room and settling themselves around the table into indiscernible chairs. Three council members joined the group after I did, lending credence to my impression of spontaneity.

Two of the last three Light Beings to enter displayed the faces of Nanci's biological parents for only a second before that illusion was replaced with their luminescent Light Being forms. My earthly parents' participation both thrilled and shocked me. Their human hosts had died years before this council meeting, so I was excited to see their faces again even for a moment. I felt they had given me the wonderful gift of recognition to reassure me in advance of the meeting about to take place. The fact that my human parents were on my council disturbed me, however, as I felt I had not respected their guidance as well as I should have while they were in human life. (I might have taken them more seriously had I known they were going to be on my council!)

The last council member to arrive appeared in human form, projected a sense of being rushed, and remained in human appearance for some time. He lofted one haunch onto the corner of the table nearest me but otherwise remained standing. When I raised my gaze to his, he looked me straight in the eye and mouthed the word "surprise!" And I was surprised! I recognized this Being as a human I know this life as Jeff, though I cannot be certain now *which* friend named Jeff. An alarm went off in my mind seeing someone from my human life, as if some emergency had triggered a recall of even earthbound Council field agents.

This second council meeting was brief and to the point. The Light Beings told me that extremely difficult times lay ahead—times that would cause me grave suffering. The council acknowledged my body's illness and weakened condition, intimating my human might not have sufficient resilience to revive. The council gave me the option of staying in human life, or, reawakening to Light Being level right then to avoid the suffering. They made it clear that if I chose to leave human life, it would not be considered a breach of contract on my part that I had not fulfilled my mission. If I stayed in my human host, they said, Nanci and I will endure constant suffering for her remaining lifetime. Despite what they showed me about the future, and my reluctance to condemn Nanci to more suffering, I chose to stay in human life. My motivation at the time was that I did not want to be a failure at the mission Source had given me. I felt a duty of unconditional love to give as much information as I can remember to all the other incarnated souls on Earth. I also did not want to miss witnessing what was going to happen in Earth's future, what I call the Third Epoch.

When I was fully inside Nanci again after making the decision to stay, I was certain the Council meeting had taken place. Yet the amnesia of human nature robbed me of the memory of whether the choice I made to return to the body was because I was willing to enduring future suffering for my mission's sake, or I wanted to live through the future transition in epochs. All I know for certain is that my choice was whether or not to "go through with it." The "it" part is unclear to me. I do not know now whether I will survive the transition to the Third Epoch. Nanci has in fact suffered physical pain and disability constantly since that Council meeting.

My 911 call shortly after the Council meeting resulted in my being admitted to the same hospital that had sent me home earlier. I found out from the ER doctor that the symptoms were caused by a life-threatening low blood sodium level. I was in the hospital for three days.

Shortly after this second council meeting, I attended my very first meeting of a local group of the International Association for Near-

Death Studies, Inc. The scheduled speaker failed to appear, so I was asked to tell my NDE story. Thus began my mission work in earnest.

Another deep afterlife experiencer who pens a far more complex meeting with council is Natalie Sudman, the author of *Application of Impossible Things: My Near-Death Experience in Iraq*. Natalie calls the meeting with Light Beings "the Gathering," rather than a council meeting. The Gathering was occasioned by Natalie's death in Iraq when an IED exploded the armored Land Cruiser she was riding in with security as part of an Iraqi police escorted convoy. Natalie was working in Iraq administering reconstruction contracts. As a result of the bombing, she suffered injuries to her right eye, wrist, and shoulder, and has two holes in her head that had to be covered with titanium patches.

The Light Being soul posing as Natalie in human life was on a mission to gather what could be broadly categorized as human cultural and political information for the Gathering. She uploaded this information to thousands of Light Beings in a complex data matrix during the Gathering/Council meeting. Near the end of the meeting, Natalie and the Light Beings agreed upon specific tasks she was to perform once returned to her former body. She reports that her agreement to perform these tasks was entirely her own choice, based on whether she thought she would enjoy them. She felt no compulsion to do this work. Natalie's body was partially healed so that it would function enough for her to fulfill the new aspects of her mission. Her report of this council meeting makes fascinating reading.

Natalie makes it clear that she and all of the personalities in the Gathering were equal. Regardless of the task each was performing, she says that none of them can properly be classified as being part of a governing body of any type. Nor can any task be considered to be the height of spiritual advancement or steps on any type of hierarchy. Although Natalie did not have any type of life review, as often occurs during an afterlife experience, she tells us that there was no judgment made of how well she had performed her mission to date. She explains: "A hierarchy of power is entirely absent, and no evaluative or punishing judgment is in evidence."[225] Natalie engaged in no self-evaluation either.

Expanded afterlife experiencer Ron Kruger is the only other soul I have found besides Natalie and myself who seems to have had a Light Being council meeting in the afterlife. When Ron was fifteen years old, he and some buddies got drunk at a small bar known to serve minors. They left the bar in a car driven by another drunk friend, driving 90 miles an hour on back roads, in a journey that ended with Ron getting thrown into the windshield, then hanging from his blood-soaked head wedged between the glass and the rear-view mirror bracket. He was extracted by the driver's friend, who awoke in the backseat and pulled Ron free without regard to whether he would sever his head in the process. Ron and his rescuer staggered away until Ron's blood loss was fatal.

Ron's next memory is of floating near the ceiling of the hospital, watching a doctor hastily stitch his wounds closed as a nurse did CPR on him. He realized he was dead and being pulled home by a wonderful force. He entered a brilliant Light after following a process he identifies as very familiar to him. Ron calls the Light the "Heavenly Plains" and describes it as an undulating force field.

Ron met with what he calls "the Council of Love," which he describes as three separate yet connected spirits radiating an unimaginable force in the form of Light. Ron was given a life review of his just ended human life, plus a newsreel-like vision of past and potential future events. The Council told him that if he returned to human life he could affect the impact, or even the outcome, of the future events he was shown. Ron resisted the mission because he saw it would involve great pain. The Council made it clear that there would be no adverse repercussions if he refused the mission and flooded him with unconditional love. Eventually, the Council persuaded Ron to return to human life but he returned with no recall of what the mission is.

These council meeting afterlife experiences lack the typical aspects of a traditional near-death or afterlife experience. This is why I placed them in a separate chapter as their own special category of expanded afterlife experiences.

16

No Manifested Heavens

SOME THRESHOLD AFTERLIFE EXPERIENCERS WHOSE accounts I have read or heard report the appearance in heaven of various earthly environments enhanced with more vivid colors or auras. According to NDE expert Dr. Bruce Greyson, his research subjects reported that heaven could not be described in words and nothing had what could be described as a physical appearance. Dr. Greyson's conclusion is that there is no typical physical image attributable to heaven. My research subjects expound on that conclusion. None of us expanded afterlife experiencers saw any physical scenery or environments while we were in the afterlife.[226] Based on my study and Dr. Greyson's, there is no objectively real earthly appearance to the afterlife. So why do some afterlife experiencers see crystal cities or libraries, or beautiful gardens, or meadows with sparkling streams? The answer is the divine power of manifesting.

Manifesting is the exercise of Source's supernatural power to create the universe that humans perceive as physical matter, according to Knowings several of my subjects received.[227] What humans believe is physical reality is actually manifested/imagined mentally by Source and its various Light Being characters. We incarnated Light Being souls harvest that power to create the individual physical lives our human hosts live.

We Souls Manifest Physical Life

Pamela K learned in the afterlife that Love, her name for the Source phenomenon, is all that actually exists and that it manifests itself in the infinite forms of Creation, including what we experience on earth. Natalie Sudman agrees, explaining that all Source consciousness cooperates to co-create and maintain what humans think of as physical world reality. As part of Source, we souls have the power to manifest and co-manifest what humans consider to be physical life. We souls manifest into our human lives what we have learned since conception to truly and deeply believe about human life and our place within it, whether it is objectively true or not.

The nature of manifesting was explained to Mira Sai, born Artie Gupta, after her body died and she looked back toward Earth and saw that nothing was there. No Earth, no galaxy, no universe. She asked mentally how it could all have just disappeared. In response, she received Knowings that the only reality is that which is permanent and changeless, that is, the Supreme Consciousness/Source which exists as a flowing presence in everything, including humans. Artie/Mira asked where the world and everything else came from. Here is her dialogue with Knowings in response:

> Like all manifestation, the world too is the creation of the great illusion or great delusion, which being the Creative aspect of this Supreme Consciousness, or the Lord, is the divine movie-projector of life … Just as a mirage in the desert disappears when viewed from a certain perspective, your life as Arti on earth has disappeared, when viewed from the perspective of the Divine Self, where you are now. Only the eternal is real; and from the viewpoint of that Reality, all that is non-eternal, disappears.
>
> I asked, "So my life as Arti never really existed at all, it was an illusion?"
>
> It replied, "oh, it existed; just as a dream exists, or a movie, or a mirage."

> I continued questioning, "So the world, was just a figment of my imagination? How did I create it? With my thoughts and desires?"
>
> The reply echoed all around me, "Y-e-s-s-s-s!" as it reverberated within my being.[228]

Chantal L returned from the afterlife with the sense that she has divine manifesting powers and that she must use them wisely. She reports:

> I no longer feel like a victim. I create my own reality and I can improve it at will. I can even influence the reality of others. My sense of power is mixed with a sense of humility. I feel a sense of responsibility towards my fellowman and towards all living creatures for I have the power to influence things into being.[229]

Chantal built a temple-like space in her basement that she uses now when she wants to focus on manifesting something into her human life. She reports a 90 percent success rate within about a month of expressing her manifesting intentions. She knows that if something she intends does not manifest, then it was not appropriate for her.

The fact that we souls can manifest physical reality for our human hosts proves we are Source, for only Source can create physical matter. Emanuele's afterlife experience confirms this:

> In fact, as God, we are all able to create our own lives according to the emotions we feel. There are no limits the wishes that we can fulfill and this too forms a part of our spiritual development. Moreover, our beliefs create our reality to the extent that we can manifest the divine on the astral plane. This always comes from within as we are the only true God.[230]

We souls inside human bodies always have the power to create physical reality for our bodies—a power that is uniquely ours as part of Source. That power is supercharged once we get into the afterlife.

Manifesting in the Afterlife

Three times while in the Light the first time, I consciously "manifested" (that is the word that entered my mind as I did it) temporary earthlike environments and scenarios that were every bit as solid and real as anything I had experienced while inside Nanci. I did it just by thinking the words "tunnel," "meadow," and "hospital corridor." The scenes disappeared as soon as I remembered I was no longer on earth. After having this profound experience, I was astounded that I could ever have been so fooled into believing earth was real. It became obvious to me that earth is actually a manifestation, as was Nanci's life.

The core entity Source informed me that the physical environments and human-like figures that threshold afterlife experiencers encounter in the early phases of the afterlife are in fact manifested appearances designed to soothe the soul's transition from human life to the afterlife. This transition process is needed to reduce the trauma of feeling the pain and fear of the soul's human host dying. The transition also reacclimates us to our natural spiritual state. These comforting transitory manifestations stem from the soul's human beliefs. Newly arrived souls manifest/create the physical environments they experience in order to match their beliefs about heaven or hell, and which of these they think they deserve.

Once freed from our cumbersome bodies, we souls instantly manifest whatever we expect to see. Many threshold afterlife experiencers expect heaven to look just like Earth, only glorified. That illusion is what they manifest/create for themselves to enjoy, perhaps seeing glorious mansions or pastoral settings. They roam beautiful gardens and hear joyous music. Some souls expect heaven to be the storehouse of Akashic Records, which is a mythical compendium of all events, thoughts, words, emotions and intent theorized by the Theosophy religion. Such believers manifest libraries full of scrolls to confirm their expectation. In other words, souls can manifest their own personal heaven to be exactly what they believed as a human to be true.

Dea M learned when she crossed over while on life support that we souls create the physical reality scenarios we experience in heaven. In her words: "At some point, I understood the essence of creativity and was given the 'joystick' of my own imagination so to speak, to create my own visions. It was amazing; whatever I thought became vision and swirled around me in its own reality."[231]

Leonard explains that everything that can be imagined can exist in our own manifested heaven because it originates from universal consciousness, what we call Source. "Thought is creative!" he exclaims. "Here [Earth] things occur with a delay, but on the other side, they happen instantly."[232]

Sadhana concurs that NDErs manifest the heavens of their dreams. She told NDE researcher Dr. Barabara Rommer that as you rise in the Light, "you can just think anything and it happens!"[233]

Newly arrived souls may also manifest human appearances for Light Beings they meet. Natalie Sudman relates to those who may call these beings "guides" or "angels," though they are not. Natalie explains that the beings are actually non-physical and consist simply as a point of energy in space. But such beings can take on any physical form they wish for a particular purpose, such as to comfort a soul that recently crossed over. The beings' appearance reflects what the soul perceiving them prefers; Natalie perceived them as vaguely humanoid because she had just been incarnated into a human. Some threshold afterlife experiencers see them as deceased friends or relatives, which they may or may not be.

The manifestations a newly arrived soul sees or hears in heaven are not real to anyone but the soul manifesting them. An individual soul's manifested reality is not shared by others. And it is not permanent. Manifested physical scenes in the afterlife last only so long as the soul needs them for emotional support while awakening to the truth that it is not human and no longer lives on Earth. Obviously, an entirely spiritual realm cannot scientifically support the weight of physical earthlike scenery. Yet, that truth does not diminish the impact and comfort a soul feels while seeing familiar surroundings.

17

Familiarity With the Afterlife

ANOTHER SURPRISING ELEMENT IN THE REPORTS of expanded afterlife experiencers is that they found the afterlife and all it holds to be familiar. Perhaps some vague memory of our origin lingers from one incarnation to the next.

Aaron M says that, "Everything seemed so familiar and unusual at the same time" inside the portal that took him to the afterlife. The feeling continued into the afterlife: "I suddenly had a feeling of familiarity and understanding that I had existed before. I felt like I had been in the place many times before. It was like a feeling of coming home."[234]

Similarly, Andy Petro tells us: "The Light was a form that I had never seen, but it was not new to me. Somehow, I knew it. The Light had a voice that I had never heard, but it was not strange to me."[235] Andy also recognized the Light's (Source's) "smile" and laugh.

Anonymous describes entering a cavern or dark place within the tunnel-like structure he entered upon getting out of body. He says that this darkness was familiar somehow.

Dea M realized she no longer had a body after leaving it lying in a hospital on life support and going into an intense yellow-gold Light. Yet being bodiless did not feel strange to her. Dea experienced vivid three-dimensional colors and musical notes in the Light, ones she had never encountered in human life, but they seemed familiar to her. Dea felt both amazement at what she was experiencing and at peace with it. To her, it felt normal.

Once he let go of the body, Doug F's consciousness opened up like a lotus flower to universal inclusion and acceptance. "This was far beyond any expectation," he writes. "Yet, at the same time familiar. Hence it wasn't overwhelming."[236]

Mira Sai describes her sensation of familiarity with transitioning to the afterlife this way: "It was the strangest feeling, and yet so familiar. It was as though I had before experienced it many times." Mira floated bodyless in a dark void before she entered the Light, but she was not afraid. Filled with the unconditional love and energy of the Light, Mira whispered to herself that she had been here before and had finally made it back.

Natalie Sudman notes in her book that the thousands of Light Beings she met with in her afterlife council meeting were familiar to her. She recognized one of them in what she calls "the Healing Environment" as one of her old friends. All nine of the beings with her in the afterlife just before she returned to human life were intimately familiar to her.

Peter N was surrounded by lights that he calls stars/conscious beings. He was familiar with them because he knew them from before he incarnated into human life. They communicated that they knew him, that they were happy to see him again, and that they loved him very, very much. Peter became aware that the entire universe is permeated with consciousness, including empty space. He notes that he was not the least bit surprised by this. He felt it was perfectly natural and felt completely safe and at home among these star beings. Peter also realized that he was familiar with the afterlife space in which he found himself: "I knew that I had been here before, so much so that I regarded this a 'home'"[237]

Ron Kruger enjoyed a simple transition from human life to the afterlife. No tunnel. No loved ones meeting him. Ron explains his familiarity with the afterlife by saying, "I knew the way well."[238]

Sue C notes in her account that she was familiar with the colors of Source's sea of life although she had never seen them before.

William Horden tells us that as he lay dying on an ambulance gurney, a deep calm came over him because he knew he was moving to an invisible but familiar place outside his body. At the hospital, the paramedics moved him to an emergency room table and put an oxygen mask over his face. Despite the mask, William's heart and breathing stopped and he was catapulted into the afterlife. "My higher soul stepped forward, speaking reassuringly about how it had been through this so many times before," he wrote.[239]

Valeska felt like she had known and interacted before with at least one of the conscious star lights orbiting Source.

I recognized returning to the afterlife as soon as I left my body and entered a calm, dark, comfortable void. I saw a pinpoint of Light in the distance and said to myself: "I know what this is. I'm supposed to go into the Light." Other aspects of the afterlife seemed familiar after the events occurred.

Bobby R was one of only two expanded afterlife experiencers in my research group who were actually told they had been in the afterlife before this visit. Bobby received the following warning from a Light Being right before his life review that what he saw would change him forever:

> Bobby, I'm sorry for the pain this meeting will cause you. When I created the Universe, I put rules and limitations in place. Every time you come here, it changes you, because this is your second time here. You will remember more than you're supposed to, and it will cause you more pain than you know. You will suffer as no one in your family ever has, and I can't change that.[240]

Bobby's life review revealed his future, which included his sister's rape and his mother's death. After returning to his body, Bobby was tortured his whole life by the anxiety of dreading when these horrible events would happen. Every time his mother got sick, he would panic, thinking that this was it—her death.

None of my research subjects explained why they recognized the afterlife. Although I too recognized each step in my afterlife journey as it happened, I could not predict what was going to happen next. I knew intuitively, however, that I had been through the process of leaving physical life and returning home many, many times.

I compared the reports of my 33 expanded afterlife experiencers in an attempt to determine whether the familiarity factor hinged on age, cause of death, length of time the body was dead, number of former physical lives, or anything else that could explain it. The data holds no answer.

18

No Barrier to Source

WHAT CATEGORIZES AN AFTERLIFE EXPERIENCE account as an expanded one, rather than a threshold afterlife experience or NDE, is that the soul encountered no barrier to reuniting with Source. Threshold afterlife experiencers may describe some type of physical or other barrier they could not pass, like a stream or bridge, or a being who prohibits them entry into the deeper stages of eternal life. Others are resuscitated fairly quickly, cutting short their afterlife visit. These spiritual adventurers return to human life without witnessing the pure energetic power of the Source phenomenon and without having the opportunity to awaken to the knowledge that they are Source itself, as expanded afterlife experiencers do.

Although I have scoured physical and online resources over thirty years, I found most of my research group subjects in the Near-Death Experience Research Foundation (NDERF)'s online database, which is the largest such database in the world. My subjects submitted their written accounts to that research organization, which then posted them at nderf.org. The submission includes a freestyle chronicle in the experiencer's own words plus answers to a lengthy questionnaire.

One of the NDERF questions asks in various words whether the experiencer encountered a border, point of no return, or barrier in the afterlife beyond which the soul was not permitted to pass. Only three answer options are offered: (1) no, (2) "I came to a conscious decision to return to life," or (3) "I came to a barrier that I was not permitted to cross; or was 'sent back' against my will."[241] Option 2 is non-responsive to the question. Technically, a voluntary decision does not constitute a border, point of no return, or barrier. Option 3 includes two answers describing different events. The soul might have encountered a physical-appearing barrier or boundary. Alternatively, a Light Being or deceased relative might have told the soul it had to return to human life, perhaps because it was not that body's time to die. Selecting option 3 can be interpreted either way.[242] Option 3 is also open to multiple interpretations of what being sent back against one's will means.

The question about a barrier or boundary appears to be based upon the reports of only threshold afterlife experiencers, all of whom returned to their bodies without personal interaction with the core Source phenomenon because the barrier or being stopped them. My research subjects, however, all returned to their bodies *after* meeting Source. Their returns were due to a conscious decision to return, resuscitation, sudden consciousness of loved ones pleading with them to return, being persuaded to return, being told they had to return, or unclear reasons. Responding to resuscitation, persuasion, and loved ones' pleas was interpreted by some in my group as being sent back against their will, causing them to select option 3. Their response could unfortunately be misread as meaning they were prevented by a barrier from having the deep afterlife experiences they obviously did have, setting up a conflict between their narratives and questionnaire answers.

My subjects encountered no manifested earthlike barrier or boundary keeping them from Source. None of them was stopped by a being from further progression into the deeper stages of afterlife. The very heart of the afterlife, and indeed life itself, is reunion with Source. I myself and all of my research subjects reunited with Source in its raw, energetic, spiritual nature. This is the ultimate and deepest afterlife experience possible. In the words of expanded afterlife experiencer DW: "We are each pieces of a greater whole.... And getting back where we can all be together again is the ultimate 'going home.'"[243] Aaron M, while inside Source, was told repeatedly by thousands of voices: "You're home now.... This is where you came from and this is where you'll end."[244]

Obviously, all of my research subjects came back from the afterlife to their human bodies. Why that happened varied from subject to subject. Some returns can be attributed to several factors. My classifications are illustrative only.

Chose to Return to Human Life

Aaron M, an atheist, was told by thousands of voices he was "now returned to the source energy ... it's time to be absorbed back into the energy."[245] Aaron repeatedly mentions in his account that he was not ready to be in the afterlife. After the last time he said it, he spontaneously returned to his body and coughed up vomit. He still felt very ill. His chest was pounding, and the room was spinning, but he was alive as a human again. Aaron answered the question about barriers with this explanation: "I came to a definite conscious decision to return to life.... I felt like I was definitely dead. Then I was provided with a chance to return, and I returned very suddenly."[246] Clearly, Aaron's wish not to be in the afterlife was honored, but it was not a barrier to his merger with Source, which had already occurred.

Anonymous met with an energetic presence radiating ultimate intelligence and a powerful force that he hesitates to call God because it so far exceeds any religious concept humans have of God. Anonymous was soaking up information via Knowings when he spontaneously, without warning, moved back through the tunnel he had traveled through to get to Source, through the universe, and back into his home. He hovered a few feet over his body, watching it die. The energy presence he assumes is God/Source was still there, giving him the instruction to "choose." He chose his family and was slammed back into his body. Even though his report confirms that he clearly made the decision himself, and encountered no boundary or barrier to Source, Anonymous chose the NDERF answer option of reaching a barrier or being sent back against his will. I count him as a voluntary returnee because he was given a clear choice to return to human life or stay in the afterlife after he had encountered Source.

Distracted From the Afterlife

Bridget F's auto accident returned her to the afterlife, where she witnessed what she perceived to be God/Source. She had no defined religious beliefs when she died and was flabbergasted to realize that God actually does exist. Like Anonymous, Bridget was receiving Knowings when she suddenly remembered that she had a body on Earth. She feared that if she did not return to her body it would cease to function. She interrupted the flow of information and "'jumped' like a salmon going upstream back to my body."[247] She clearly decided on her own, and for her own reasons, to come back. Bridget's questionnaire answers contradict her narrative report. She responded to one question that her experience included a boundary, which it did not. The version of the NDERF questionnaire that Bridget answered asked whether she reached "a border or limiting physical structure." She responded, "No. My body would die." Then, in the next question about a border or point of no return, Bridget selected the answer of coming to a barrier she was not permitted to cross or was sent back against her will. Regardless of her questionnaire answers, Bridget did not hit a barrier and was not prevented from meeting Source according to her own words. She was distracted away from living in the afterlife by thoughts of human life.

Pamela K understood that she had the choice of whether to stay on the "other side" or return to what she calls the game of being an imaginary personality. That realization caused her to turn toward the Light. She was delighted to be home and filled with joy and bliss. She realized that she is and always has been the "I Am" that she considers to be God/Source. After reveling in this bliss, Pamela had the sudden thought that it was not fair that she was having such a wonderful experience without her family. That shift of focus to human life caused her to zip at warp speed back into her body. Pamela responded to the NDERF questionnaire that she was sent back against her will even though she was simply distracted from the Light by thoughts of her family. The return occurred after she had fully enjoyed the bliss of being Source. She reached no barrier to the Creator.

Sue C, like some other expanded afterlife experiencers, moved swiftly through blackness toward a rainbow of colors. Sue pushed herself mentally to get closer to it and then, she says, "I had broken through to this other side." To the extent that Sue describes a border, she crossed it and reunited with Source. The rainbow of colors pulled her into its "soul shaking ecstasy."[248] She was one with Source. Sue perceived a resistance that she interpreted to be a boundary, but not a physical one, according to her NDERF questionnaire answers. Source told Sue that she could not return to where she was, on Earth, if she stayed within it. Suddenly memories of her home and family flooded her mind. She missed her boyfriend. Source took this to be her choice and Sue sadly backtracked and awoke being resuscitated by medical staff.

Returned to be Missionaries

As soon as she remembered in the afterlife that she had come to Earth to accomplish a mission, Chantal L was asked mentally by a Light Being whether she wanted to stay or return to her body. She immediately responded that she had to go back to finish her mission. She soon found herself back in her former body. Notwithstanding the fact that it was her own decision to return, Chantal answered the NDERF questionnaire saying that she had come to a barrier she was not permitted to cross or was sent back against her will.

Ron Kruger was given a choice of whether to return to human life by the council monitoring his already existing mission. He was told that he could dramatically impact the outcome of future earth events if he returned. But, he was told that whatever he chose would not diminish the Council's love for him. Ron reluctantly chose to return and willed himself back into his body. He responded to the questionnaire by saying he made a conscious decision to return to human life.

After her extensive council meeting and healing sessions, Natalie Sudman simply popped back into her body to continue the mission she had started before birth. That is a spontaneous return.

I had just awakened within Source to the knowledge that I am Source playing a temporary role as Nanci when I started the return to my body. I was not told it was not my time. Rather, my distinct impression all along had been that it was Nanci's time to die and my time to return to my home. I obviously did not reach any type of barrier or boundary because I had been literally inside Source at the time I started returning. Instead, I felt I had been given a mission back on Earth. Once I received the answers to my most important spiritual and existential questions, I was convinced that someone had to tell the other souls incarnated into humans the truth about life, death, the afterlife, and Source. I certainly did not intend for that spokesman to be me. Yet, here I was, the designated messenger, whirling back into Nanci's body against my will. I do not know whether I actually volunteered for this mission or just the desire for there to be a messenger propelled me into that role.

Mira Sai's body died in an auto accident. She reports that while in the afterlife she became aware that Source is her true identity. Mira suddenly had a faint memory of the world of separateness after she had merged completely with Source. She felt anxious about that glimpse of individuality and asked Source where she was supposed to go from there. To her dismay, the answer she got was that she had to go back and do the "Work," which is what she calls her spiritual mission. Mira answered no to the NDERF question about whether she reached a border or barrier or was forced to go back.

Had to Be Persuaded to Return

When a Light Being asked Doug F to return to his body, he refused. Next, he saw the persona of a man sitting at a café table, inviting him to sit as well. The persona introduced himself and told Doug that Source needed him to go back. Doug felt he could not refuse this request a second time, but he did condition it upon being able to retain the memories of his afterlife experience. Doug answered the NDERF question about his return with the option indicating that he had made a conscious decision to come back even though he did not want to do it.

Leonard reports that Source told him he had to go back. But he refused to return to his sick body. Source showed him a vision of how upset his mother was about his death. He agreed then to come back, reentering his former body through the head. Interestingly, Leonard describes the process as "like putting a diver suit too small on."[249] It felt the same way when I reentered Nanci's body. Like many of my other research subjects, Leonard responded to the questionnaire that he reached a barrier or was forced back against his will even though he had been persuaded to return.

Pulled Back by Family Pleas

Immediately after she could feel herself touching the Creator, Dea M spontaneously started regaining consciousness. She fought against the return, but her family was trying desperately to bring her out of her coma. Dea writes: "I did not want to return to this reality. But it was almost like a door shutting and I was back in my own body."[250] Dea answered the NDERF questionnaire with the option that stated she had reached a barrier she was not permitted to cross or was sent back against her will. Her narrative indicates that she did not want to return to human life, but there is no indication that she was sent back by anyone. More important, Dea's family's pleas did not prevent her from meeting Source because she heard those pleas after reunion with Source.

Joan LH had already merged into Source when she started to awaken from her coma. Her younger brother was yelling at her to come back. Joan believes she returned voluntarily in order to one day be the mother of two wonderful daughters whom she believes will change the world. She responded to the NDERF questionnaire that she did not reach a border or point of no return. Joan was one of several who seemed to return to her body after loved ones called her back from blissful reunion with Source.

Kathryn H tells us that she joined with an ocean of souls in merging their individual selves into Source. She nevertheless returned to her body for two reasons: she felt her family calling her back, and she wanted to complete her spiritual goals on earth. Kathryn reports she was reluctant to come back but felt it would be too painful for her family if she did not. Once again, this expanded afterlife experiencer marked the NDERF questionnaire answer of barrier/sent back against her will even though she chose to return from the afterlife for her own reasons. Moreover, she made her choice after she had fully merged into Source. Her return posed no barrier to that experience.

Robyn rejoined Source during her expanded afterlife experience, realizing at once that her, and everyone else's, true nature is as Source. Robyn later became aware she was back in her body when she heard her name being called very loudly. She took a deep breath, and she was back. She answered the NDERF question by saying that she did not reach a border or point of no return.

Not Their Time to Die

Andy Petro was told by the Light after his body drowned that he had to return to human life. He refused several times. After the last time he was told this, Andy was *instantly* back on the beach in his body being resuscitated by one of his friends who pushed the water out of his lungs. Although Andy responded No to the question whether he had reached a barrier or was sent back against his will, I count Andy as having returned because it was not his time to die. Andy did encounter Source in its raw nature and learned the eternal truths about life. He reported that he became aware that he is the Light, what he calls Source. He describes being absorbed into and becoming an indistinguishable aspect of Source. He states that he *is* Source, as we all are. This reunification with Source proves Andy did not reach a barrier to it.

Demi B drowned at age fourteen and was reunited with Source while in the afterlife. Demi realized that she was God/Source when she felt at one with its energy, that she herself was the energy source of all love and wisdom. She was thereafter given a message that it was not her time to come home and that she had things she needed to do back in human life. Demi answered the NDERF question about boundaries by stating that she was sent back against her will. The admonition that it was not her time obviously did not prevent her merger with Source.

DW and what she calls the Being of Divine Love were in communion, with DW receiving Knowings in response to all her questions, when Source suddenly told her she had to go back because it was not her time. She responded to the NDERF question by choosing the option of being sent back against her will. Her return occurred after reunification with Source.

Yazmine S had just started to feel "the Great Presence (which 'they' call God) pervade my very core, as if my entirety is exploding into love," when a powerful voice declared: "It is not your time."[251] She mentally begged to stay but awoke in the hospital two days later. She answered the question of whether she came to a barrier she was not permitted to cross, or was sent back against her will, in the affirmative. Yet, her forced return came after she had not only encountered Source but merged into it.

Wayne H understood while he was in the Light that he was going to return to his body, even though he wanted to stay in the afterlife. Wayne witnessed Source's raw beauty and oneness, learning that there is no individual but only the One [Collective] Source. He discovered that he was not to join Source at that time and felt a sensation of two choices, one peaceful and comfortable and the other cold and dark and wrong. Without knowing he had made his choice, Wayne felt the Light darkening. He fell back through blackness into the light surrounding the gurney his body was lying on. Wayne reports that he was unconscious for thirty hours after that, underwent surgery, and died again without returning to the afterlife. Wayne answered the NDERF questionnaire by choosing the option that referred to reaching a barrier he was not permitted to cross/being sent back against his will. I count him as returning because it was not his time to die based on his foreknowledge that he would be returning to his body.

Resuscitation

Peter N was completely merged into God/Source/the Light to the point that he actually was the Light. Afterward, once he felt he had regained some individuality, Peter wondered to himself where he would go next. He decided to go into an area of stars separated by darkness and started to move in that direction. Suddenly, he became aware of someone calling his name and opened his physical eyes to see that he was in an ambulance. He responded to the NDERF questionnaire that he had not reached a barrier or been sent back against his will.

Malinda K describes in her narrative that she not only saw and was drawn to Source, but she also recognized that she is Source herself. Malinda was revived by her husband performing CPR, so she responded to the NDERF questionnaire that she was brought back against her will. I count her return as resuscitation because her narrative has no description of any return decision making. The resuscitation did not present a barrier to Source or any type of afterlife boundary.

Sudden, Spontaneous Returns

Anke Evertz, who accidentally set herself on fire, died in a Munich, Germany hospital during the course of a nine-day coma. Anke reports that she was dissolved into what she calls the Source during her expanded afterlife experience. Near the end of her afterlife visit, Anke was asked by her Light Being companion if she knew where she was. She responded that she was home, at the Source, where we all come from. The Being said that it would show her where she was and pulled her back into her body. Once there, Anke recognized Source in every cell of her human body, something she had never noticed before. The Light Being told her that she is the embodiment of Creation. Anke did not submit her account to NDERF, and therefore did not fill out the NDERF questionnaire. I did not count Anke as having returned against her will, though she did not make a conscious decision to return. Her account makes it sound as though she was suddenly transported back to her body after completing her reunion with Source.

Bobby R remembers merging with the Light/God to become one. He was born legally blind but was able to see clearly in the afterlife. Bobby reports that he did not see any border or boundary beyond which he could not pass. Clearly, he was not prevented from accessing Source. He was in the midst of the future aspects of his life review when he suddenly woke up in his body, confused and overwhelmed. His answer to the NDERF question about a border or being returned against his will was no.

Henry W witnessed small golden orbs of Light joined into what he believed to be God/Source. While Knowings were downloading into his mind he realized that he would not be staying in the afterlife. After all his questions were answered, Henry felt himself falling backwards and landed with a jerk back into his body. Although his return was spontaneous, Henry answered the question about hitting a barrier or being sent back against his will in the affirmative. No barrier prevented his access to Source.

Unknown Reasons for Return

Emanuele and Gwen J do not explain how or why they returned to their bodies. They both report in the questionnaire that they encountered no barrier or boundary. Both experienced Source and learned that we are all God/Source.

Jennifer J died four times and had different experiences each time. She does not mention how or why she returned from her one expanded afterlife experience. She does, however, clearly describe her reawakening to the knowledge that she is Source, as we all are. Therefore, whatever propelled her back to her body did not present a barrier to Source for her.

Sadhana's expanded afterlife experience included returning to what she calls the Godhead and becoming completely dissolved within it. Her brief account does not describe her return to the body other than to say that by the time she returned her body had been taken to the hospital from her house. Sadhana's dissolution within Source proves she encountered no barrier to reaching it. Sadhana did not submit her account to NDERF.

Valeska K answered the NDERF questionnaire saying she did not come to a border or point of no return. Her narrative gives no explanation but does record that she witnessed what she understood to be Source. Whatever caused her return to her body did not pose a barrier to Source.

William Horden merged with Source in what he describes as being like every drop in the ocean and suddenly becoming aware that he is the ocean itself. He answered no to questions about whether he reached a boundary or limiting physical structure, or a border or point of no return the second time he submitted his expanded afterlife experience account. He answered yes to those same questions in his first submission. Based on his written account, it is clear that William was not prevented from reaching Source and becoming One with it.

William W answered no, he did not come to a border or point of no return in the NDERF questionnaire. His body drowned in the ocean at a California beach when he was a child. After observing that Source is a pure love energy composed of everything, including us, William returned to his body. His body popped up about thirty yards from where he entered the ocean. He encountered no barrier to Source.

Regardless of how or why they returned, these expanded afterlife experiencers have given us a huge gift by bringing back with them the wonderful, loving, eternal truths they received from Source in the afterlife. They also provide evidence that it is possible to return to human life after meeting and/or merging with the Source phenomenon. Expanded afterlife experiencers passed well beyond the core NDE elements accepted by medical NDE researchers.

19

Abandoned Religions

THE MAJORITY OF MY RESEARCH SUBJECTS changed their religious beliefs as a result of their experiences with the Source phenomenon and its Knowings they gained in the afterlife. Most of the expanded afterlife experiencers in my research group who were Catholic or Christian before they died abandoned those religions. Those who were Buddhist did not. Disbelievers in God became believers in Source.

Christians

Andy Petro regarded himself as strict Roman Catholic before he entered the afterlife. Now, he is no longer religious and considers himself to be "a 'spiritual being' who believes that we are all one in the Light with God."[252]

Anonymous went from being a non-practicing Catholic to having no religious affiliation though he regards his expanded afterlife experience as religious.

Doug F was a Catholic before meeting Source and does not engage in belief systems now.

Chantal L explains that she was a practicing Mormon before she died but is now affiliated with no church. She did attend Sunday School a month after her experience and heard the priest tell those present that the Lord sends us trials to test our faith. Chantal's whole being reacted strongly with sadness to this statement. She thought to herself: "No one who has ever been to the other side could believe such a thing."[253] She never went back.

DW had been a non-practicing Protestant before her suicide and did not consider religion to be important in her life. Since returning from intimate union with Source, DW chooses not to attend a specific church. Rather, she supports places and people who care for others, regardless of their faith. She says that her beliefs no longer fit within any religious denomination, noting "This [the afterlife experience] was totally outside my religious beliefs and learning.... For me it was a deeply spiritual experience that required no 'faith' because I have seen it and experienced it."[254]

Joan LH was not only a devout Catholic before her body's death at age twenty-five, she also held a bachelor's degree in theology and worked as a youth minister in a Catholic parish. After returning from Source, Joan left the Catholic Church and affiliated with a Unity Church and the Movement of Spiritual Inner Awareness. Joan tried very hard to reconcile her experience with Catholic dogma, but she now knows for a fact that "each of US is part of God, and God is far more than any one of us. There is no 'heaven' and no 'hell'–except what we do to ourselves."[255] It became impossible for Joan to reconcile her beliefs before death with her knowledge and experience in the afterlife.

Jennifer J was reared as a Catholic but attended a Lutheran Church at the time of her body's death. Her religious beliefs were replaced by the Knowings she received in the afterlife. She says that prior to her four NDEs, she believed in purgatory, heaven, and hell as places you go to up there somewhere to spend all of eternity. She learned none of this is true. She writes: "When I came back from the experience, the words 'they have it all wrong' resonated within me when thinking about major western religions." Jennifer adds: "I have an entirely different understanding of God after the experience. I believe that all are related, and trying to pin it all down into one religion or another is missing the point."[256]

Bobby R came from a home where his mother was Protestant and his father Catholic. Bobby was only four when his body died. He now considers himself as having more in common with the Christian Seeker movement but focuses his religious practice on creating a relationship and not worship. Bobby writes: "I saw God; but God didn't care if I prayed, if I went to church, God only wanted a relationship and wanted me to be happy and know love."[257]

Demi B, who was around fourteen when her body died, called herself a Protestant before her expanded afterlife experience and is now unaffiliated with any religion. She explains that she had beliefs as a child but received Knowings in the afterlife. "A belief is a belief," she writes, "knowledge is power."[258]

Henry W responded to the NDERF questionnaire about faith before he died with the choices of both conservative/fundamentalist, and none, without explaining the contradiction. Now, he says he is a liberal or New Age Christian. Like Henry W, Ron K answered the NDERF questionnaire about religion by saying that his religion changed from conservative/fundamentalist to liberal after his council meeting in the afterlife.

Malinda K was a non-practicing moderate American Baptist before she entered the afterlife and identifies now as either liberal or an atheist. Malinda explains the change: "I no longer believe in a separate omnipotent God character. I believe we are all 'Gods,' all connected. There is no 'heaven' or 'hell.' We just are."[259]

Wayne H was a Southern Baptist before he met Source and received Knowings in the afterlife. Now he says he has very little use for religions.

William W's beliefs changed dramatically due to his time in the afterlife. Before then, he was a Protestant. Afterward, he stopped going to church because he knew the teachings were not accurate. He admonishes us: "GOD is not to be feared and God does not get involved in our lives. We have within us to live a life of heaven or hell. It's our choice."[260]

Peter N was distinctly anti-Christian as a rebellious teenager before his body's fatal accident. He did not understand at that time that the eclectic spirituality he practices now could be a separate belief system. Though not a churchgoer himself, Peter was brought up in a Christian culture. He reports that no Christian teaching can even come close to what he experienced in the afterlife. Peter has some choice words for how religions sometimes misuse NDE accounts:

> I do currently feel significant disappointment with the way in which some members of the Christian faith make attempts to assess NDEs as supporting the dogmatic view that "only through Jesus" will anyone be "saved," "go to heaven" or "have an afterlife worth having." To me that is a scare tactic to attempt to force people to accept Christianity as the ONLY way. … It's a damaging proclamation and far removed from my experience. I can see why someone that has an NDE with Christian motifs and/or characters might see things in that way…. But when it comes to churches promoting that, … it is an abuse of NDEs. … I would also say the same for any religion trying to commandeer NDE accounts as proof of their own religious dogmas.[261]

Robyn was raised Christian but left that faith at age thirteen to pursue Buddhism and Wicca. After her expanded afterlife experience as an adult, Robyn is spiritual and accepts of all forms of belief. She has a warning for us all about organized religions: "Religion is a tool that is used to drive a wedge between people. There is far more monetary profit in separating people than in uniting them. The higher-ups continue this game of instilling fear to dominate because they believe power and money will make them feel better."[262] Robyn notes that she thinks the less a soul knows about religion the better it can understand the truth about Source.

When she was five years old, Yazmine S attended Church of England Sunday school and had visions of Jesus and angels throughout her childhood. She wanted to be a nun when she grew up. Yazmine did not really have a belief system at the time her body died. After her expanded afterlife experience, Yazmine searched for a religious affiliation that fit with her own experience of Source. The closest she has come is Tibetan Buddhism.

Before I died the first time, and for a few years thereafter, I was a sometimes practicing Catholic. I had had twelve years of Catholic education and six years of Methodist college before going into the afterlife upended my whole life. Like Joan LH, I tried desperately hard to reconcile my former beliefs with the Knowings I received and experiences I had while deep inside the afterlife. I could not. I eventually wrote *BACKWARDS Beliefs: Revealing Eternal Truths Hidden in Religions* in an effort to at least create some peace for myself. I feel that beliefs are no longer relevant to me because I now have knowledge and, with it, power over my own life.

Buddhist or Hindu

Gwen J was Catholic, Buddhist, and Taoist before she died and added Hinduism to her eclectic beliefs after she returned to human life.

Kathryn H had been Buddhist for forty years before her expanded afterlife experience and remains so today.

Mira Sai was born Hindu, educated in Catholic convent schools, and practiced Tao meditation as an adult. After her afterlife visit, Mira says she is spiritual rather than religious.

No Religion Identified

Aaron M called himself an atheist before he encountered Source and received Knowings deep inside the afterlife. Afterward, he says he is more spiritual but not religious. He notes: "I have developed a belief in something greater than us."[263]

Sue C had no religious affiliation before she died, referring to herself as an atheist. Now, she says she does not get wrapped up in religion because "God is the fabric of existence, the rest does not matter."[264]

Emanuele was agnostic before visiting the afterlife and considers his religious affiliation to be that of liberal faiths now. He responds to the questions about religion with: "Now I know for certain that every one of us is God and that eventually everyone will realize this."[265]

Bridget F states that she was not baptized and had no formal religion prior to her death. When asked if her religious practices have changed since her experiencer, Bridget pulls no punches and exclaims: "Yes. G-d is love and anything else is crap."[266]

Leonard likewise practiced no religion either before or after his expanded afterlife experience.

Dea M selected the term "moderate" for her religious beliefs before she died, with no explanation of what that means. To describe her belief system after experiencing Source, she selected the NDERF questionnaire multiple choice response of "moderate to no religion." She explains that there were no religious aspects to her expanded afterlife experience, and that God just is.

All of these former members of diverse religious sects discovered that religions are based on human ideas rather than the truth about Source and life itself. They were all able to relinquish their former divisive beliefs and accept that we are all exactly the same, equally powerful, and joyously united in one entity—Source. How wonderful it would be on Earth if all people could open their hearts and minds to this unifying knowledge.

20

A New Category of NDE Emerges

I HAVE ALWAYS KNOWN THAT NDE RESEARCH IS INcomplete because I lived all of the afterlife phases described in the medical literature and NDE accounts, and then some. I remained in heaven longer and progressed through far more wonderful and mind-blowing later phases of eternal life than the traditional NDEr or threshold afterlife experiencer. I can tell from what other expanded afterlife experiencers write that they too have surpassed the experiences of most NDErs. I have created a new category of NDE for us: the expanded afterlife experiencer.

Evidence Before This Study

Near-death experience medical researchers, utilizing patient data supporting hundreds of scientific papers and books published over nearly fifty years, have proven beyond a doubt that consciousness survives death. This body of research establishes that a *near-death experience* (NDE) is a bona fide medical phenomenon shared by an estimated 850 million people worldwide.[267] Medical researchers use the term near-death experience to describe evidence of consciousness leaving the body when a person is close to death, or when the body has died temporarily and later revived. Those of us with religious or spiritual backgrounds call the consciousness that leaves the body the "soul." Regardless of your preferred term, medical research has established that during an NDE consciousness remains intact and capable of having experiences despite its separation from the human it previously inhabited.

The term NDE has been broadened over the years to include events and experiences that provide insight not only into death and the dying process, but also into the afterlife. Recent studies on afterlife experiences provide partial insights into the Creator. A brief review of the history of NDE research shows how what was originally thought to be a simple physical phenomenon has become a highly complex metaphysical one.

The Original NDE Concept

The term "near-death experience" was coined in 1975 by teaching psychiatrist and author Raymond Moody, Jr., in his paradigm-shifting first book, *Life After Life.*[268] This book of NDE accounts ushered in a new field of medical research. Raymond Moody earned a PhD in philosophy from the University of Virginia and began his career teaching philosophy. He later graduated with an MD degree from the Medical College of Georgia, planning to teach philosophy at the medical school level. Dr. Moody wrote the 1975 original edition of *Life After Life* while still a medical student. It launched his parallel writing career and became a best-seller.

Life After Life describes fifteen possible NDE elements that Dr. Moody culled from interviews of 150 NDErs and clinical death survivors. Dr. Moody's NDE concept delved well beyond simple near-death physical conditions. Those elements he identified that relate directly to what happens *after* death are:

- Consciousness gets out of the human body;
- Feelings of peace, quiet, and a lack of pain;
- Unusual auditory sensations (like buzzing, ringing, music);
- Entering a dark space (tunnel-like experience);
- Becoming aware of nearby spiritual beings (deceased loved ones);

- Encountering a Being of Light that emanates love and warmth;
- Undergoing a life review;
- Hitting a border or limit beyond which the soul cannot pass.

Each of these elements, except for getting out of body, describes something humans can experience while in the body and therefore accept comfortably. All of us have felt moments of peace, quiet, and lack of pain. We all have heard buzzing, ringing, or musical sounds, especially if we have a smartphone. Anyone who has traveled much has been through a dark tunnel. Concepts of ghosts, apparitions, and even deathbed visitors and other spiritual beings such as angels are familiar to most. Common parlance includes the phrase "light at the end of the tunnel" when speaking of death and everyone expects love and warmth at the tunnel's other end. Similarly, the idea of life events flashing before one's eyes at death is widespread. We all have encountered barriers or borders of one type or another, even to the point of anticipating one in the afterlife when it is not our time to die. None of these original NDE elements even hints at an event that is otherworldly or beyond human comprehension.

Dr. Moody included an event in his core NDE elements if some or many patients reported it. He found no account where a patient reported all fifteen of his elements. None of the patients described in *Life After Life* progressed beyond the life review, which I know from personal experience comes early in the afterlife. I call an experience after death that includes having a life review a "threshold afterlife experience" to distinguish it from the broader expanded afterlife experience. A threshold afterlife experience includes the core elements of consciousness leaving the body and passing through some type of change in state of existence, commonly termed "going into the Light"; becoming aware of a new state of awareness and existence, often called heaven or the afterlife; receiving knowledge telepathically; often seeing one or more spiritual beings; and perhaps having a life review before returning to the physical body.

Measuring the "Depth" of an NDE

Some of Dr. Moody's NDErs and death survivors were not only near death, but their souls had some threshold afterlife experiences. These are the patients who had spiritual life reviews—as opposed to events simply flashing before their physical eyes. I believe Dr. Moody would call these few instances "deep" or "complete" NDEs because he uses those terms in his book. His list of elements does not include the types of events my expanded afterlife experience research subjects experienced. I suspect his patients failed to reach these additional experiences because they perceived themselves as hitting a boundary or limit in the afterlife, consistent with Dr. Moody's last core NDE element.

Kenneth Ring, PhD, a social psychologist and former professor at the University of Connecticut, is another pioneer in the NDE research field. Dr. Ring followed up on Dr. Moody's deep or complete NDE concept by conducting his own scientifically structured study to determine what constitutes a "deep" NDE. He created measurements to categorize experiences by depth. Dr. Ring described five temporal stages of an NDE as (1) feelings of peace and contentment; (2) detachment from the body; (3) entering a dark area of transition; (4) seeing the Light; and (5) entering through the Light into another realm of existence. The first two stages occur frequently without death or any chance that the soul is "crossing over," as we say. These two stages can be experienced by getting out of body, without being close to death. The third and fourth stages of darkness followed by Light mark the border between simply being out of body and starting the journey into the afterlife. The last stage in Dr. Ring's description applies only to what I would characterize as afterlife experiences in the initial stage, or threshold afterlife experiences, rather than to all NDEs in general. Many reports labeled as NDEs describe only getting out of body and staying in the physical world with no afterlife events.

Dr. Ring's 1980 book, *Life At Death: A Scientific Investigation of the Near-Death Experience*,[269] is based on interviews conducted in 1977 of over one hundred people who came to the brink of death, or, in a few cases, actually returned from clinical death.[270] Dr. Ring and his associates compiled a core NDE experience index, still used today, which weights each component in an attempt to measure depth of the experience scientifically. Dr. Ring weighted entering the Light as the most important aspect of a core NDE, followed by seeing or hearing a spiritual being, and having a life review. These are actually afterlife events, not physical dying or near-death experience events. However, it is possible for a human to experience an earthlike version of each of these elements without dying or crossing over. They are therefore easily accepted by science as part of a medical NDE rather than as spiritual phenomena.

Dr. Ring's core NDE elements do not include the soul reaching a physical border or limit, as mentioned by Dr. Moody. Rather, his research showed that the majority of core experiencers came to a decision point about whether to continue the passage toward permanent death or return to their body. Dr. Ring's interviewees reported two methods of decision-making. The first method happens when a deceased loved one tells a soul in the afterlife that it is not time for its body to die. Dr. Ring explains that the loved one advises the soul it has a decision to make and strongly encourages it to return to physical life. The second method occurs when a Light Being or other presence informs the soul that it must make a decision about whether to stay or return, sometimes showing the soul a life review that might include the future as part of the process. Once again, Dr. Ring's data show NDErs being prevented from entering deeper or more expanded afterlife experiences because they encountered this decision point and chose to return to their bodies.

Another NDE research pioneer, psychiatrist Bruce Greyson, MD, met Raymond Moody at the University of Virginia when he supervised him through an internship rotation in psychiatric emergency services. During this residency, Dr. Moody's book *Life After Life* gained worldwide recognition when a major New York publishing house reissued it. In response to overwhelming interest in his book, Dr. Moody scheduled a meeting of like minds at the University of Virginia. Drs. Ring and Greyson attended, among others. As a result, Dr. Ring, Dr. Greyson, cardiologist/NDE researcher Michael Sabom, MD, and social worker John Audette co-founded what was later named the International Association for Near-Death Studies (IANDS), a non-profit organization dedicated to fostering research into NDEs.

As an IANDS member, I became aware of the "Greyson Scale" for measuring the depth of an individual's NDE. In 1983, Dr. Greyson published his own scale for measuring the depth of NDEs in an article titled "The Near-Death Experience Scale: Construction, Reliability, and Validity."[271] The Greyson Scale provides a sixteen-element checklist based on interviews with patients who collectively experienced seventy-four NDEs. Not surprisingly, psychiatrist Greyson's scale elements focus less on the events experienced during an NDE and more on the experiencer's physical and emotional perceptions. An NDEr's responses to the sixteen items are weighted and counted in determining the depth of the NDE. Many of the Scale elements describe events unrelated to going into the afterlife. The elements of the Greyson Scale that do relate directly to what happens in the afterlife are:

- Did you view scenes from your past? [Life review.]
- Did you suddenly feel like you understood everything? [Receiving Knowings.]
- Did you see or feel a bright Light?
- Were your senses more vivid than in human life?
- Were you aware of things going on in other locations on Earth? [Out-of-body travel.]
- Did you see scenes from the future?
- Did you meet a mystical being or presence? [A Light Being.]
- Did you see dead people or religious spirits?
- Were you stopped from proceeding further by a border or boundary?[272]

An experience that meets only a few elements of the Greyson Scale of NDE depth is sufficient to qualify as an NDE for medical research purposes. Dr. Greyson's depth measurements include even more afterlife events than Dr. Ring's and moves the NDE research model further away from the study of physical dying conditions. NDE research thus becomes more metaphysical in nature.

Over the years, Dr. Greyson has published over a hundred scholarly articles on his NDE research. In his 2021 book, *After*, he mentions that his group of research subjects included many hospitalized patients as well as over a thousand NDErs who have sent him their written accounts. Although the Greyson Scale includes more of the elements of an expanded afterlife experience than previous studies, it still assumes the soul will reach an impediment to further progress and to meeting the Source.

The ground-breaking research of Drs. Moody, Ring, and Greyson, as well as others who have studied the depth of NDE experiences, inspired my research. Their work convinced me that the types of experiences a soul lives in the afterlife depend upon the depth of the afterlife experience itself. This became the foundational hypothesis for my research.

The NDE Elements Are Expanded

Cardiologist Sam Parnia, MD, of the New York University School of Medicine, reported in 2021 that he uses a more extensive NDE core element list of 50. Most of the elements appear to parse Dr. Moody's original 15 elements into smaller segments. Dr. Parnia writes: "These [50] themes were derived using samples containing 200—500 survivors of life-threatening disorders using quantitative and qualitative methods and were verified using a survey of over 6000 people from an international poll carried out across seven different countries, with diverse cultural and religious backgrounds."[273] Dr. Parnia's elements include one (no. 38) that refers to the Source, the origin, but no discussion of this element is provided. The Parnia expanded list of elements continues to include the soul reaching a barrier beyond which it cannot pass, thus prohibiting it from accessing the origin, the Source.

Radiation oncologist Jeffrey Long, MD, likewise an NDE researcher, made NDE reports available to the general public in 1998 via his Near-Death Experience Research Foundation website (nderf.org). His goal was to collect as many NDE accounts as he could that included answers to a detailed questionnaire he devised to examine the individual elements of the phenomenon. In 2010, Dr. Long published his first book, *Evidence of the Afterlife: The Science of Near-Death Experiences.*[274] In it, he analyzes 613 sequential submissions to NDERF's website to prove through nine lines of evidence that there is life after death.

Dr. Long's questionnaire includes NDE elements of a spiritual nature. This field of research thus moved wholeheartedly into the subjects previously dominated by religions and spiritual belief systems. The NDERF questionnaire seeks information about God, love, the afterlife, the purpose of life, religion, a comparison of the experiencer's beliefs before and after the NDE, and how the experience affected the person upon returning to the body. Nevertheless, Dr. Long, like his predecessors, focused his official findings in 2010 on the traditional core elements of the NDE identified by his predecessors, adding only the elements of intense and generally positive emotions or feelings, and a sense of alteration of time or space. Obviously, both of these new elements are common in human life. Dr. Long continued to use the core element of whether a soul encountered a boundary or barrier to a more fulsome afterlife experience.

Within a few years, the NDERF online database of NDE stories grew by leaps and bounds to over four thousand submissions. As he reviewed the questionnaire answers as they came in, Dr. Long noticed that around two hundred of them mentioned perceiving God in some way. That prompted a new research project. In 2016, Dr. Long published the results of his "God Study," a review of all NDE accounts submitted to nderf.org between November 11, 2011, and November 7, 2014. First he identified those that met his previously established core NDE criteria. He then reviewed the NDErs' responses to multiple-choice questions, as well as narrative responses to open-ended questions in his NDERF questionnaire, for use of the word God or G-d. He found 277 that met his study criteria. His findings are published in *God and the Afterlife: the Groundbreaking New Evidence for God and Near-Death Experience.*[275] Dr. Long's book includes NDEr reports on subjects not previously discussed in medical NDE research, including the receipt of Knowings about the universe, whether there is judgment and/or punishment in the afterlife, and the purpose of life. For the first time, the God Study describes threshold afterlife experiencers' perceptions of God.

In a paper submitted to the Bigelow Institute for Consciousness Studies in November 2021, Dr. Long listed as both current and state-of-the-art the same elements comprising the traditional core NDE elements, which are getting out of body, heightened senses, positive emotions, going through a tunnel, encountering the Light, meeting deceased loved ones or mystical beings, a sense of alteration in time or space, a life review, encountering otherworldly realms, learning special knowledge, hitting a barrier, and return to the body. In my parlance, these are elements of the threshold afterlife experience and stop short of the expanded afterlife experience. This stance is obviously more limited than Dr. Parnia's 50-element core NDE.

The Added Value of My Study

The foregoing NDE research pioneers paved the way for this novel study of expanded afterlife experiences. Their work inspired me to delve extensively into finding more afterlife experiencers like me who were not barred from meeting, and merging into, what I call Source, the Creator of the universe. I sought to prove that the depth of an afterlife visit determined whether a soul experienced Source's raw nature and power, understood the origin of souls, and learned Source's reasons for creating them.

The accepted core medical NDE elements describe experiences and perceptions familiar to most humans. None of the elements describes a supernatural or non-human event, or something unfathomable to those who have religious or spiritual beliefs. The core elements include a requirement that the soul encounters a border or barrier beyond which it cannot pass. This requirement severely limits medical NDE research findings to reflect only the data of souls who were prevented from direct union with Source.

My study takes the next step in NDE research by analyzing afterlife reports in which the soul did not encounter any type of barrier, boundary, border, or other bar to a full-on experience of the deity entity/Source in its raw form, stripped of all human preconceived notions.

I call an account that describes passing beyond a demarcation of the afterlife state of existence (such as going into the Light) an "afterlife experience" rather than an NDE. I find it to be more precise. Afterlife experiences occur at later stages than nearly all NDEs, many of which are limited to out of body events that take place on the physical plane. An afterlife experience account documents the nature of, and life within, the afterlife itself. Afterlife experiences transpire when the soul lives events beyond those traditionally accounted for in most near-death and death survivor out-of-body accounts.

I coined the term "threshold afterlife experience" for those accounts that describe events up to and including having a life review (see chapter 8) and/or the experiencer receiving Knowings (described in chapter 11) before returning to human life. In a threshold afterlife experience, the soul is confronted by some type of barrier to further progression toward the types of experiences my study group has reported.

"Expanded afterlife experience" is another term I created to distinguish this research study group members from threshold experiencers. I abandoned Dr. Moody's terminology of a "deep" NDE because the word implies that some afterlife experiences are superficial when they are all profoundly life-altering. Although each soul's afterlife perceptions are specific to that soul, there are established events that a soul may or may not participate in during its journey. Like the common elements in NDEs, there are common elements in expanded afterlife experiences, as described in chapters of this book.

While medical researchers may occasionally recount excerpts from NDE stories that include elements unknown in human life, they decline to delve more deeply into them. This makes total sense to a medical researcher, for as Dr. Greyson says in his book *After*, "Questions like 'Where does consciousness go after the body dies?' seem to be pushing the limits of scientific research beyond the breaking point."[276] Others, like Dr. Moody, deny that there can ever be scientific proof of life after death, claiming:

> In fact, scientism–the doctrine that scientific method is the only way of establishing knowledge–is a red herring in this instance. The question of the afterlife, the deepest of mysteries, is not ripe for scientific inquiry. And the proper, scientific attitude toward life after death is to acknowledge that it is not yet a scientific question. Rather, it is a conceptual, philosophical question, and a very difficult one at that.[277]

Dr. Long's scientific research, however, has already opened the door to answering the question of where consciousness goes after death. My research blows that door off by explaining how and why our "individual" portions of consciousness exist in the first place, as well as what happens to us after death.

Souls who have left human life behind for the afterlife unanimously agree that there exist no words in any human language that come anywhere close to accurately describing the happenings there. No vocabulary exists for afterlife experience because it is nothing like human life. The best afterlife experiencers can hope for is that rough analogies, metaphors, similes, and vague references to things within human understanding convey the gist of what we mean. We know our words are not accurate. Yet, we are compelled to try to share our stories as best we can.

Unfortunately, nearly all NDE researchers use those imperfect words to categorize and dissect our stories. They count how many of us use the term "tunnel" leading into the Light. How many of us use terms describing an otherworldly realm? How many mention a barrier? How many use the word God? And more. They count words and phrases to calculate percentages of souls who experienced certain things without regard to the fact that there is no common vocabulary among experiencers. I know from trying that categorizing and quantifying afterlife experiences based on words used is akin to defining beauty by counting the number of times each color is used in paintings in the Louvre. Nearly all of those scientific researchers counting the words we use to describe the afterlife are not themselves near-death or afterlife experiencers. I am. I have had such deep and rich afterlife adventures that I can look back with hindsight and recognize what was actually happening from the word pictures NDErs describe in their reports. I do not have to rely on their imperfect word choices.

My research project combines the best of both worlds. I followed scientific principles, and I included research subjects who had experiences far outside those of human life.

Existing Near-death Experience Categories

Based upon my analysis of roughly two thousand firsthand reports, and NDE medical literature review, I identify the following categories of NDEs in ascending order of completeness and complexity.

Out-of-Body Experiences

A person does not have to be dying in order for the soul to leave the body and return to it later. Out-of-body (OOB) experiences may also occur while the body is in a medical condition that could, or does, lead to death. These events are literally "near-death" in time but not necessarily the result of death. In my estimation, many NDEs reported in the literature and online are of this nature. The soul gets out of the body and moves around in physical space but never begins the process of "crossing over" to the afterlife.

The OOB category also includes the accounts of patients who left their body and hovered above it shortly after an injury or while in the hospital. A few souls reported being at the ceiling of a hospital room. A couple of physicians, including NDE researcher Sam Parnia, MD, took reports of ceiling-hovering literally and conducted research projects to study how many cardiac arrest patients reported seeing signs and symbols pasted near or on the hospital ceiling. I know from my own OOB experiences in hospitals that a patient may report being on the ceiling merely because the soul is judging the distance at which it hovers to be about the height of a room ceiling. Rarely does an NDE account in this category describe the ceiling composition or its light fixtures, which would indicate that it was literally at the ceiling. A group of NDErs explained to Dr. Bruce Greyson why these experiments did not produce many NDE accounts: "Why...would patients whose hearts had just stopped and who are being resuscitated—patients who were stunned by their unexpected separation from their bodies—go looking around the hospital room for a hidden image that has no relevance to them?"[278]

Many NDErs who meet the medical definition of having a near-death experience (see below) get out of body but do not begin the process of crossing over to the afterlife. They are literally near death but not post-death. They are still in the incarnation stage of eternal life and remain on earth. Their accounts are thrilling and life-affirming. They give us hope and reassurance that life does not end when the body dies. In my opinion, out-of-body experiences prove beyond doubt that we are not dependent upon a human body for our consciousness, awareness, or personality. However, these engrossing stories tell us nothing at all about the afterlife because these NDErs did not enter it. As a result, these discarnate souls have no opportunity to rise above human thinking and perceiving. They stay in the physical world, though disconnected from it, and continue to think and believe as humans do.[279]

"Crossing Over" Process Experiences

I define "crossing over" as the process through which a soul progresses when returning to spiritual life after the body's death (temporary or permanent). Crossing over starts when the soul leaves the physical matter state of existence. This category includes those accounts where the soul returned to its body after it explored a dark space, outer space, a void, a tunnel/vortex/doorway/portal or in any other way left the physical world but did not go into the Light. Some of these accounts are fascinating in their vivid detail of tunnels and travel through the universe. Ultimately, however, they tell us nothing about the afterlife because the soul did not get that far.

"Entering the Light" Experiences

Based on research and my several personal experiences entering the afterlife, I conclude that a soul must move through some type of transition or passage to a different state of existence in order to access what we call heaven or the afterlife. Many NDE accounts of this nature specifically mention going into the Light and describe it. Others do not mention going into the Light but the passage they do relate clearly transpired within the afterlife. For me, this category of NDE includes all those accounts where the soul briefly entered the Light and had some experiences but hit a barrier or was told it could go no further. Dr. Kenneth Ring's research discloses that among those whose progress in the afterlife was halted, "Virtually all experiencers feel that either they themselves decided to 'come back' or that they were 'sent back' (in a few instances, apparently, against their own preferences)."[280] I include entering the Light experiences in my category of afterlife experiences, even though many of these souls just barely made it into heaven. These are *threshold afterlife experiences* in my parlance.

Threshold Afterlife Experiences

A "threshold afterlife experience" is my own terminology. It occurs when a soul leaves a body at its death, crosses over into a different state of existence, continues on into what religions call heaven, has limited experience there, and returns to its former body. In order for an afterlife experience to exist, three events must occur: (1) the soul must enter a different state of existence; (2) it must have conscious experiences of what the soul believes to be heaven or the afterlife before returning to the body; and (3) it includes Dr. Moody's barrier point or Dr. Ring's decision point that prevents the soul from reunion with Source.

The oral and written reports of these courageous souls who entered the afterlife form the core database of NDE afterlife research. Without their testimony we would have nothing but speculation about what awaits us on the other side. These experiencers' willingness to share their most intimate moments after death gifts us all with knowledge and comfort to literally last a lifetime. Yet, most of the NDErs who begin the afterlife experience witness only parts of it before they are halted by a being or physical-appearing barrier from progressing to reunion with the Creator. This barrier gives them no more than a brief glimpse into the afterlife from the perspective of standing on the threshold of it. But these souls have no idea that they are merely on the threshold of a far, far greater divine Source phenomenon. To them, the afterlife holds nothing more than what they experienced at the threshold. Their perceptions are naturally couched in commonly understood religious and spiritual terms and analogies because generally nothing foreign to their religious beliefs happens to them.

Threshold afterlife experiences often include a life review and receiving Knowings on various topics. They often meet Dr. Raymond Moody's NDE criteria. Most afterlife experiences reported are of this nature, but not all NDEs are.

A New Category of NDE is Identified

"Expanded afterlife experiences" include some elements of the threshold experience, such as receiving Knowings and the life review, but add many others.

The vast majority of threshold afterlife experience accounts reinforce the impression that heaven is a place, calling to mind the physical places we have on Earth. Many see entirely real-looking earthlike scenery (the explanation for this is in chapter 16 ("No Manifested Heavens")). Expanded afterlife experiencers, on the other hand, realize that what humans call heaven is a state of existence rather than a physical place. Thus, we do not experience the afterlife as an earthlike environment.

Expanded afterlife experiencers continue through a transitional stage and reach later eternal life phases before returning to their former bodies. Their accounts give us an understanding of life free from the human interpretations offered by religions or science—or by threshold afterlife experiencers.

The states of awareness, or stages of eternal life, that I and others achieved beyond the typical NDE criteria place human life into context within the infinity of eternal life. This context leads to the inescapable conclusion that consciousness survives the death of a human body because human incarnation is just a tiny part of eternal life. It holds no more and no less significance than any other incarnation into physical matter.

Science Has Validated NDE Accounts as Reliable

A principle of the scientific method as applied to NDE research, according to Dr. Jeffrey Long in *Evidence of the Afterlife*, is: "In reaching conclusions about these [NDE] accounts, we followed a basic scientific principle: *What is real is consistently seen among many different observations.*"[281] My data demonstrate a consistency among expanded afterlife experience accounts that shows they are real. In other words, the same traditional NDE research principles that validate other NDEs also validate the more complete and fulsome accounts of those souls who have traveled far beyond the NDE accounts upon which medical research is customarily based.

The validity of after death experience accounts has been verified by the results of tests used by various scientists and studies conducted by medical NDE researchers.

What are those tests? Dr. Bruce Greyson's review of the scientific research convinced him that NDEs are very different from hallucinations and do not correlate with mental illness.[282] Dr. Karl Jansen, a neuroscientist who for years insisted NDEs were caused by drugs or brain chemistry, eventually admitted that these things could be a trigger for having an NDE but did not actually cause them.[283] By repeatedly interviewing the same NDErs over decades, and finding no differences in their stories over time, Dr. Greyson established that "experiencers' memories of their NDEs are reliable."[284] He and another researcher compared NDE accounts reported before the publication of *Life After Life* with those reported after that book's contents became widely known to see whether NDErs were just recounting what they expected to experience based on Dr. Moody's well-known research. The result was that NDE descriptions have not changed over the decades and do not simply parrot familiar concepts of what we assume happens at death.[285] To further test the reliability of NDErs' accounts, Dr. Greyson and collaborators conducted an experiment designed to determine whether the memories of NDEs were of real events or imaginary ones. The experiment proved not only that NDErs' memories of their experiences were of real events but also that the NDE events were remembered as even more real than human life events.[286]

In addition, Dr. Jeffrey Long documented nine lines of evidence that prove the validity of his research subjects' reports of life after death. His book *Evidence of the Afterlife* demonstrates why logically and medically NDEs are credible memories of real experiences. With respect to the validity of encounters with God, Dr. Long explains in his 2016 book *God and the Afterlife*, "This is not religious dogma or theology.... In other words, people are not merely stating or projecting their religious yearnings or beliefs.... The fact that they describe these encounters so similarly gives us confidence that they have, indeed, met the same Being."[287]

Dr. Sam Parnia approached the validity of NDEs from the opposite direction. His research concludes: "Little evidence supports the notion that the recalled experiences of death, including paradoxical episodes of lucidity with conscious visual and auditory awareness can be categorized as hallucinatory, illusory, or delusional."[288]

Internationally respected cardiologist and NDE researcher Dr. Pim van Lommel, along with others, conducted a prospective study of cardiac arrest patients to obtain more reliable data that would either corroborate or refute existing theories on the cause and content of NDEs. "We were particularly surprised to find no medical explanation for the occurrence of an NDE."[289] Dr. van Lommel's findings were published in the widely respected British medical journal *The Lancet* in 2021.

If medical research proves the reliability and reality of NDErs' memories of human-like experiences while out of body, it also proves the reliability of expanded afterlife experiencers' memories of a supernatural Source and events. After all, these take place within the same type of experience. Nevertheless, some NDE researchers still hesitate to venture into the realm of the supernatural. Dr. Bruce Greyson explains his own reluctance: "almost 90 percent of them (NDErs) say they encountered some kind of divine or godlike being. That posed a problem for me, because I couldn't think of a way to test the accuracy of any of their accounts.... My ingrained skepticism holds me back from taking these descriptions literally."[290] Fortunately, Dr. Jeffrey Long did not share that hesitation when he published his own God Study. Importantly, he validated his subjects' impressions of God using the scientific method, concluding:

> An essential scientific principle is that what is real is consistently observed. In a previous section we found that when near-death experiencers see ongoing earthly events while out-of-body, they are generally exceptionally accurate. Given this, it is reasonable to believe that when NDErs consistently describe their often unearthly spiritual experiences, they are similarly accurate.[291]

It is time we applied the same research-based conclusion that what threshold afterlife experiencers see and learn is reliable, and give credence to the non-human-like aspects of the afterlife that expanded afterlife experiencers see and learn.

Implications of My Study

A subgroup of NDE accounts exists that documents the experiences of souls who were not blocked from accessing the Creator and who learned the origin of souls and their relationship to the Creator.

I have labeled these souls expanded afterlife experiencers because they went far beyond the scientific research category of deep NDE. Consistent reports from the research subjects evidence the existence of a Creator that is non-humanoid, non-physical, and more analogous to a scientific phenomenon than a being. Many subjects felt the name "God" to be inappropriate as too limiting and some adopted the term "Source" for the Creator. I refer to the Creator as Source as well because it is the source of existence.

Subjects learned the divine Source phenomenon is All-encompassing. It is collectively composed of a core identity plus all of what we call souls plus the entire universe of physical matter. Most subjects learned that Source created the universe to provide opportunities for its soul parts to gather experiences otherwise unavailable to the core identity due to its intrinsic nature. Nearly all subjects learned everything about the universe, including information previously unknown to humankind. All of them lived events of a supernatural nature.

Two unique characteristics of this subgroup were identified. First, none of them encountered a barrier, boundary, being who refused them further progress, or other impediment as is identified by scientific researchers as a required core element of a deep NDE. Second, nearly half of the subgroup reported familiarity with one or more aspects of the afterlife and Source.

My study proves that evidence does exist for the reality of a divine Creator deity, sometimes called Source. The evidence is unbiased by religious or spiritual beliefs because of the diversity of the subjects' pre-death beliefs. It is consistent among many different types of experiencers who died for a variety of reasons. It further provides evidence of how and why human, and all physical, consciousness was generated by Source. It answers the question of what the purpose of life is and what goals we souls seek to achieve. Ultimately, the study expands our concepts of God and the relationship between God and humans previously based on ancient, primitive human-centric speculation to a deeper understanding of Source/God, the origin of souls within Source, and the relationship between Source and souls incarnated into humans that is now based on scientific study.

21

Research Method and Findings

THE PURPOSE OF THIS STUDY WAS TO DETERMINE whether a subgroup of deep afterlife experiencers could be identified from among roughly two thousand near-death experience (NDE) accounts. The target subjects had to provide objective data about personally meeting the Creator/deity and receiving information about the intrinsic nature of the deity, the nature of souls, and the relationship between the two. The goal was to ascertain whether there was a particular trait or aspect of this subgroup of NDErs, as opposed to all others, that allowed them to have personal contact with the Creator/God and obtain this information.

Study Design and Participants

Afterlife experiences are a small percentage of near-death experiences, a field of study of widely researched and verified medical phenomena. An afterlife experience occurs when a soul leaves a body at his/her death, "crosses over" into what religions call heaven, has various experiences there, and later returns to its former body.

There are many categories of NDEs, not all of which are acknowledged by medical researchers. Some people consider religious ecstasy experiences to be NDEs. I do not. Others include spiritual experiences during extreme emotional trauma to be NDEs. Again,

I do not. A few refer to "empathic NDEs" or "shared NDEs" where another person witnesses a dying person's out-of-body or dying experience.[292] I had an experience like this with my mother while holding her hand as she died. I watched her leave her body and enter a dark area. That frightened her so much she turned back and traveled around and around her hospice room out-of-body, pausing briefly in front of the priest, who confirmed the event. I consider my mother's experience to be an NDE just prior to permanent death. I do not consider my own observation or that of the priest who also witnessed it to be an NDE.

Medical personnel rightly consider the temporary death of a person to be a near-death experience whether they have any memory of a spiritual nature or not. My research does not include any case lacking an afterlife visit. There are several other types of soul events labeled as near-death experiences, including episodes where the soul goes through part of the crossing over process or simply gets out of body, that do not involve activities in the afterlife. These types of events likewise are not part of my research project.

The near-death experience accounts describing events indicating that the soul did enter the afterlife, and, possibly had a preliminary life review before being resuscitated or stopped from further progress, are what I call "threshold afterlife experiences." These souls stick their noses into the afterlife tent, so to speak, but go no farther. My research does not include these types of reports, though they are clearly afterlife experiences. Many of these experiencers lived several elements of what I describe as expanded afterlife experiences. Many have received similar Knowings to those described in this book. It was difficult at times to decide whether to include an account in my group of expanded afterlife experiencers. The cutting edge was whether the account was consistent with those quoted in chapters 1 to 3 of this book.

An expanded afterlife experience is the most extreme form of afterlife experience. The primary screening tool I applied to determine what is and is not an expanded afterlife experience is that the account must describe witnessing, or even uniting with, the Creator/Source phenomenon. I did not include accounts based solely on the author stating that they believed they met or merged with God. Often these souls mistook the Light to be the Creator itself. The Light is absolutely part of Source's vast corona, but it is not the core Source phenomenon. Nor did I include accounts stating the soul met God in being form, such as a Light Being or human religious persona. These being-like appearances are manifestations created by the soul itself, perhaps with the aid of Light Being helpers. The account was included only if the author describes reuniting with a deity that has no religious or spiritual belief system characteristics. In other words, the experiencer must have met (and possibly became) a non-humanoid deity—a divine phenomenon in its raw form and nature without the projected appearance of a human god (see chapters 1—3).

The defining criteria of an expanded afterlife experience are: (1) the soul was not prevented by a barrier, boundary, or being from reuniting with its Creator/Source; (2) the soul witnessed the magnificence of the infinite Source phenomenon without any projected humanoid form or religious expectation; (3) the soul recognized or remembered Source's collective nature; (4) the soul became aware that souls incarnated into physical matter are literally parts of Source's own consciousness and self-awareness. In other words, the soul received the answers to: what is God? What am I? In addition, my criteria for an expanded afterlife experience includes (5) evidence of events the soul can participate in only in the afterlife, such as:

- mentally receiving downloads of Knowings about the afterlife and eternal life that do not conform to traditional human religious or spiritual beliefs (not simply knowing about it from reading materials, like this book);
- having a life review that allows the soul to feel what the people affected by its human host's behavior felt and thought;
- living other souls' physical lives through merger into their energy;
- meeting with a council of Light Beings who monitor the soul's earthbound mission;

- experiencing other non-human events unknown to those who have not entered deeply into the afterlife.

Procedures

During the six years it took me to obtain college degrees in chemistry, biology, and psychology, I often conducted scientific experiments to produce data that either confirmed or disproved a given hypothesis. I also devoted time to researching and reviewing published articles on topics in the fields of chemistry, anatomy, physiology, and psychology and wrote research papers of my own posing a particular theory or reporting a research conclusion. A few years later, I earned a doctorate degree in law and began practicing law in a 270-attorney law firm. As an attorney, I spent 36 years researching both law and medicine in order to advise my health care provider clients, and, to defend them in malpractice or healthcare fraud proceedings. This research included reviewing published medical journals and text books, as well as interviewing various medical experts.

With this background, I have experience following the protocol of creating a hypothesis, conducting research to collect data on that hypothesis, and drawing conclusions from the data. This is the scientific method I followed in compiling, collating, and analyzing the data in this expanded afterlife experience study.

Other NDE researchers selected a sample of NDE accounts, identified the ones in the sample that met their core NDE criteria, and then analyzed those to see what they might reveal. I reviewed NDE accounts for those that met my above-stated research criteria. My goal was to find other souls like me who intimately experienced Source as a non-human entity that defies religious and spiritual dogma.

I analyzed roughly 2,000 accounts labeled as near-death experiences, more than half of which were written reports on nderf.org, the website hosted by the Near-Death Experience Research Foundation. The rest were published by the International Association of Near-death Studies in its publications or websites, other NDE websites, YouTube, or are described in books written by experiencers or NDE researchers. Several of my readers assisted me by sending me links to accounts they thought might meet my project criteria. I selected only afterlife reports that could be accessed by my readers. I searched nderf.org for submitted accounts of those experiencers I found elsewhere. I analyzed each account vis-à-vis a list of 43 elements that I have gleaned from NDE research papers and books, as well as from my own afterlife experiences.

Among all of those reports, I found the accounts of 33 experiencers (including mine) that met my definition of an expanded afterlife experience. We each reported the raw data we perceived rather than anything that matches our own pre-death religious expectations. Unlike other NDE researchers, I eliminated all accounts describing a barrier, border, boundary, being, or other impediment preventing the soul from having direct interaction with Source's raw nature and power. The existence of that barrier event identifies an afterlife experience as *threshold.* The experiencers in my sample did eventually return to human life, of course. However, they were not prevented from knowing Source intimately before doing so, unlike threshold afterlife experiencers who were prevented from reuniting with Source. To make it into my research sample, the account had to demonstrate that the experiencer crossed over into the afterlife, passed the point at which threshold afterlife experiencers encounter a barrier to further progress, and encountered and understood the true essence of the Source.

Original Hypothesis–Depth of Afterlife Experience Dictated Nature

The work of the NDE pioneers discussed above, and that of many others, convinced me that afterlife experiences did in fact have varying depths. My original title for this book was *Deep Inside the Afterlife.* I was most impressed by Dr. Moody's theory that the types of events a soul experienced depended upon whether there was evidence of the body's clinical death or the length of time he/she was dead. Dr. Moody theorized:

> How far into the hypothetical complete experience [all 15 elements] a dying person gets seems to depend on whether or not the person actually underwent an apparent clinical death, and if so, on how long he was in this state. In general, persons who were "dead" seem to report more florid, complete experiences than those who only came close to death, and those who were "dead" for a longer period go deeper than those who were "dead" for a shorter time.[293]

Armed with Dr. Moody's hypothesis, I expected to find evidence within my own research group that clinical death or death for a long time explained why they had had expanded rather than threshold afterlife experiences. The first line of inquiry, commensurate with Dr. Moody's theory, was to ascertain whether the experiencer's report offered third-party confirmation of clinical death to determine if that was the key factor. Most of the reports in my group did not state one way or the other. I next reviewed the length of time the body was dead. Those times ranged from two or three minutes to, in my case, approximately half an hour dead. Unfortunately, most authors did not state the length of time they were dead. There was insufficient data to establish any correlation between the presence of confirmation of clinical death, or length of time the body was dead, and whether the soul had an expanded afterlife experience. Yet, every subject did in fact have an expanded afterlife experience.

I began looking for other reasons. The reports were far from complete in terms of providing background data. When it was available, I could find no physical reason for having an expanded afterlife experience rather than an NDE or threshold experience. No age, gender, or medical correlation. No religious beliefs reason. No spiritual beliefs reason. No death scene similarity. No cause of death reason. No correlation between expanded afterlife experience and resuscitation versus spontaneous return to the body. No clear explanation of why some souls went deep enough into the afterlife to meet Source and others did not.

What I learned from this analysis is that trying to categorize afterlife experiences by depth is futile. I also learned from Source, via Knowings I received during my own expanded afterlife experiences, that NDEs and afterlife experiences are all different because each soul's life is unique. Each life is special. Each part of Source that incarnates has different personality and character traits and physical life history. Each soul has its own distinctive perceptions based upon its eternal personality within Source, as well as from its many incarnated lifetimes. The result is that each soul's afterlife experience is also unique and special. Each is to be cherished rather than measured.

Final Hypothesis – Expanded Afterlife Experiencers Quickly Relinquished Human Thinking

My final hypothesis became that souls who had expanded afterlife experiences did so because they quickly relinquished the idea that they were still human and stopped thinking like a human. I originally believed the transition from thinking like a human to understanding as our natural spiritual selves occurs during the life review because the soul is able to literally get inside others they had interacted with in physical life to see and feel events from their perspective (see chapter 8). This theory was disproven by the number of study subjects who did not have the more complete, deeper life review. However, my research disclosed some glaring differences between the expanded and threshold afterlife experiences that indicate my subjects had escaped the human mindset and beliefs.

Outcomes

First, all of the expanded afterlife experiencers met and recognized the deity phenomenon called Source up close and personal. They were given Knowings or remembered that we souls are all Source ourselves. This is explained in great detail in chapters 1 to 3. Second, none of the expanded afterlife experiencers saw human relatives that had predeceased them greeting them at the so-called gates of heaven or later. Often, a threshold experiencer will see a relative welcoming them or later telling them to return to their bodies. Not so in an expanded afterlife experience. Third, none of my research subjects saw heaven as having earthlike scenery or anything else earthlike, as many threshold experiencers do. Fourth, none of my subjects saw a religious figure of any type and had no inclination to interpret any being they saw as a religious icon. Some threshold experiencers interpret Source's outer corona of Light to be God. Or they think a Light Being they meet is God or Jesus or another religious figure. Fifth, every expanded afterlife experiencer commented that time does not exist in the afterlife, at least in the way humans know it. In other words, time is not a universal constant as humans believe. And, sixth, none of the expanded afterlife experiencers learned that the purpose of life has anything to do with learning human-oriented lessons, learning how to love, or any of the other purposes of life popular with religion or spiritual philosophies.

Another interesting commonality is that many in my research group mentioned that the afterlife and/or the process of going into it felt familiar to them. They recognized something about it. Their reports give the impression that these souls have crossed over so many times that recognition of parts of the process stuck with them even in subsequent incarnations, a lingering memory of sorts. Perhaps this is why expanded afterlife experiencers do not need the creature comforts of earthly environments, relatives, or religious icons to feel at home.

Data Analysis

Afterlife experiencers of both types courageously share their afterlife adventures and Knowings with us. They freely admit they have to use ill-fitting religious or spiritual terms because there are no human words that can describe what is essentially a non-human experience. As NDEr Bill Urfer vividly explains: "Try to draw an odor using crayons. You can't even begin to try, no matter how many crayons you have in your box. That's what it's like describing NDEs with words."[294] Dr. Raymond Moody further expounds on this difficulty as he encountered it in his research: "We know that ineffability, or indescribability, is a hallmark of spiritual and mystical experiences. That is, people say that there are no words that adequately describe what takes place during their experiences."[295] Yet, words are all we have to work with.

Most of my 33 expanded afterlife experiencers have a written account in the nderf.org database. The NDERF submission format offers an opportunity to provide a written narrative of the experiencer's story along with answering a list of questions soliciting a yes or no answer, a multiple-choice answer, and/or a comment to explain the submitter's perceptions. Sometimes I determined that the multiple-choice answers do not adequately address the expanded afterlife experience, resulting in a loss of clarity. Also, the wording of the questionnaire changed over time and perfect alignment of responses cannot be guaranteed. A few times it was obvious that the submitter did not understand certain questions and gave answers that conflict with his/her written narrative.

The Near-Death Experience Research Foundation was unable to provide me with contact information so that I could interview my research subjects myself and obtain additional or clarifying information. My study is consequently based entirely upon the available published written or recorded material.

The following statistics reflect the number of times I perceived an element of my research parameters appearing in a verbal or written expanded afterlife experiencer report. I used my own extensive, multiple experiences in the afterlife to determine what the report describes rather than relying upon the exact words used.

One cannot assume an event did not happen just because it is not mentioned in an account. We have no way of knowing whether it did not occur or was simply not included in the account at that moment. Consequently, calculating percentages of afterlife experiencers who report various events or perceptions has very little scientific value, in my opinion. Nevertheless, I have included my totals for what insight they provide.

1. Age at time of death: 5 children, 28 adults

The NDERF questionnaire does not ask age, making it impossible to determine this element with accuracy. Based upon the circumstances of their deaths, it appeared to me that most of my subjects were adults. Andy Petro was eighteen when his body drowned. Bobby R was four years old at physical death. Demi B drowned at human age fourteen. Ron Kruger was fifteen when his body died as a result of injuries sustained in a head-on auto collision with a tree. William W was a child when his body drowned.

2. Race: not disclosed

3. Gender: 20 females, 13 males

4. Religion at time of death:

Christian 18, Buddhist 3, Atheist or agnostic 4, Hindu 1, Taoist 1. The rest did not indicate any religion.

5. Knew they were dying: 55%

6. Knew it was their time to die: 15%

A couple of subjects were told by Light Beings to let go of human life. My Light Being eternal friends told me it was my time to come home. Others simply knew.

7. Cause of death:

Causes of death included medication overdose (3), drowning (3), burned (1), anaphylaxis (2), severe asthma attack (1), traffic accident (8), sepsis (1), heart attack (3), suicide (2), bled out (2), grand mal seizure (1), blown up by IED (1), severe apnea (1), dysentery (1), medical shock (1), and non-auto accident head injury (2).

8. Length of time dead:

Most research subjects did not report how long they were dead. Jennifer J's body was pronounced clinically dead in the hospital and her family was notified of her passing. Joan LH's body was dead 5 to 8 minutes. Kathryn H's body stopped breathing for 3 to 5 minutes. Malinda K's body's heart and breathing stopped for 10 minutes. William Horden's body was dead for 2 minutes. Yazmine S's body was dead 6 minutes. According to my hospital records, my body was dead for close to 30 minutes.

9. Consciousness got out of body: 100%

10. Feelings of peace, quiet, and a lack of pain: 100%

11. Entered blackness or a void: 48%

For Robyn, the void area was bluish purple. Malinda K labeled what she experienced as deep space.

12. Had a tunnel-like experience: 36%

Aaron M did not pass through a tunnel in the physical sense. Rather, he reports that a dark shape enveloped him until he was swallowed by it. Andy Petro was underwater with his body stuck in mud when he entered a deep dark tunnel with bright Light at the end. Anonymous was transported through a 5-by-4-foot static space with a moving grey border with a small bright, soothing Light at the end. Demi B saw a tunnel of colors with beautiful music. Henry W crossed over through a lighted doorway. Leonard was sucked through a long, dark tunnel with a glowing pinpoint of light at the end and was joined by other entities traveling like him. Peter N provides a delightfully detailed description of the space he moved up and away from his body through, complete with dramatic lights and sounds. Chantal L, Mira Sai, and I returned to our bodies through something like a tunnel, with mine feeling like a whirlwind.

13. Traveled out of body: 22%

Anke Evertz visited Munich. Emanuele and Jennifer J visited friends or family. I traveled through our solar system and Yasmine traveled around the universe. Aaron M saw people on the other side of the world.

14. Saw deceased loved ones: 0%

No expanded afterlife experiencer was met when crossing over or at heaven's gate, so to speak, by deceased loved ones or relatives. Threshold afterlife experiencers often report seeing deceased relatives in the afterlife. That is why this core medical research NDE element exists.

Later in the expanded afterlife experience, a few of my subjects saw human-like apparitions for reasons other than being welcomed into heaven. Aaron M saw a friend's mother he knew to be deceased. Her job was to try to gently convince Aaron to return to his body after all Light Being attempts at logic failed. Doug F met a human-appearing man whose job was also to convince Doug to return to human life. I saw a friend who appeared in human form at my second afterlife meeting with the council monitoring my mission. Also, two Light Beings in attendance momentarily assumed the visages of Nanci's parents. Leonard answered the NDERF questionnaire by saying he saw deceased family friends, but his narrative does not mention it. I found several conflicts between a submitter's narrative and answers to the NDERF questionnaire among my subjects.

15. Saw the Light: 91%

Gwen J's experience took place entirely in darkness though she describes the darkness as including all light as well. Joan LH did not perceive a Light and says her experience took place in "Loving." Natalie Sudman saw no Light and was blinked from her mangled vehicle directly to a council meeting. I also did not see a Light the two times my body died and I went directly to council meetings. I did enter the Light the first time my body died.

My research group described "The Light" as white, silver, yellow gold, deep purple, or as spanning a full spectrum of colors.

16. Had unusual auditory sensations: 12%

Dea M heard 3-D music using notes unheard of in human life. Demi B was surrounded by musical notes she had never heard before. Joan LH heard music unlike anything humans can produce, and it felt like the music was literally a part of her. Peter N heard many different sounds while he was traveling through a tube of light and vibration toward the afterlife.

17. Mentioned not having a human body: 45%

Chapter 7 ("No Bodies in the Afterlife") details the perceptions of those in the research group who mentioned anything about their appearance in the afterlife.

18. Realized was same person bodyless: 21%

Chapter 7 ("No Bodies in the Afterlife") identifies this as a separate realization from being aware of no longer having a body.

19. Realized human life is not real: 34%

See chapter 10 ("Human Life Isn't Real").

20. Learned no time exists in the afterlife: 100%

21. Recognized the afterlife: 42%

See chapter 17 ("Familiarity With Afterlife").

22. Described heaven as physical place: 0%

Chapter 16 ("No Manifested Heavens") compares threshold and expanded afterlife experiencer accounts on this distinguishing point.

23. Became aware of spiritual beings: 67%

Dr. Moody's subjects appeared to have met one Being of Light who emanated love and warmth and was the source of the Light they saw after crossing over. Other threshold afterlife experiencers mention seeing or meeting one or more Light Beings. Various expanded afterlife experiencers describe seeing or feeling multiple to zillions of entities as Beings of Light, pinpoints of Light, unearthly beings, the presence of God, someone who felt like Christ, orbs or spheres of aware Light, and conscious stars. Many subjects reported that the Light Beings were familiar to them.

24. Had a life review: 58%

Chapter 8 ("Expanded Life Reviews") describes the two types of life reviews expanded afterlife experiencers had.

25. Learned there is no judgment: 33%

Chapter 8 ("Expanded Life Reviews") also explains that the life review did not include judgments from anyone other than perhaps the soul itself.

26. Saw past lives: 27%

Some subjects viewed and/or remembered what humans call past lives either during the life review or separate from it. Two others, not counted here, knew they were being shown past lives but chose not to watch. See chapter 9 ("Other Physical Lives).

27. Remembered Creation of the universe: 33%

See chapter 4 ("Creation of the Universe").

28. Saw Earth's past: 12%

Chapter 12 ("Earth's Past and Future") provides descriptions of what was seen about Earth's past.

29. Saw their own or Earth's future: 48%

See chapter 12 ("Earth's Past and Future").

30. Received Knowings in general: 100%

31. Received Knowings about everything: 91%

Chapter 11 ("Received Knowings") provides details about the topics of general and specific Knowings received. A few subjects who did not fill out an NDERF questionnaire did not mention one way or the other whether they learned everything about everything.

32. Received Knowings about God/Source: 100%

See chapters 1 ("We Met God in Its Raw Nature"), 2 ("We Knew God's Collective Nature") and 3 ("We Learned We Are God").

33. Met God/Source's raw essence: 100%

Detailed descriptions are provided in chapters 1 and 2. In contrast, threshold afterlife experiencers sometimes assume the Light or a Light Being, with or without a manifested humanoid appearance, is God. Experience has taught my research group that newly arrived souls may perceive Source's Energy field as Light. Research subjects sometimes thought this as well when they first encountered the Light. Then they progressed past that impression to realize that the Light is a manifestation of Source's Energy, not its core identity.

34. Learned souls are part of God/Source: 100%
See chapter 3 ("We Learned We Are God").

35. Received Knowings about the purpose of incarnation: 64%
Chapter 6 ("Purpose of Incarnation") provides the research subjects' own words about their understanding of the Knowings they received about the purpose of incarnated life.

36. Remembered Creation of souls: 24%
See chapter 5 ("Creation of Souls").

37. Saw aliens in the afterlife: 6%
Aaron M saw alien animals. Anonymous met an alien machine with a soul. I saw my own past lives as various types of alien creatures and things.

38. Learned about or experienced living lives without incarnating: 15%
Chapter 13 ("Living Other Souls' Lives Vicariously") explains the divine phenomenon of living another soul's physical lifetimes without incarnating oneself.

39. Barrier or being stopped progress: 0%
Chapter 18 ("No Barrier to Source") explains how this traditional NDE research core element divides souls into threshold or expanded afterlife experiencer categories. Its use as a required element forces exclusion of the expanded afterlife experience accounts from traditional NDE medical research. In my opinion, it skews NDE research findings toward experiences full of projected human religious or spiritual expectations, manifestations, and interpretations.

40. On a mission in human life: 33%

See chapter 14 ("On a Mission"). A couple of my subjects had missions before they died. Others gained a new mission after their expanded afterlife experience.

41. Met with a council that monitored their mission: 10%

See chapter 15 ("Light Being Council Meetings") for a detailed description of this different, and far rarer category of expanded afterlife experience.

42. Gained new abilities: 48%

The new abilities available to them in human life generally relate to paranormal activities, like being psychic, medically intuitive, or clairvoyant.

43. Changed religion or spiritual belief system after return: 82%

Chapter 19 ("We Abandoned Our Religions") compares my subjects' change in belief systems after returning from communion with Source.

Conclusions

My data prove the accuracy of my second working hypothesis: that expanded afterlife experiences happen when the soul is able to quickly shed its human perceptions and ways of thinking in the afterlife, and it is able to perceive and experience events as they are in their raw or natural spiritual essence. Forty-two percent of my research subjects had an advantage because they recognized the afterlife as events unfolded. Some of them commented that they knew the way well or had been through the process many times. Others simply noted that various aspects were familiar. They felt they were home.

Regardless of whether an expanded afterlife experiencer mentioned familiarity with the afterlife in his/her report, certain indicia support the conclusion that he/she was familiar with the process. No subject was met by predeceased friends or family members. None manifested earthlike physical scenery as their version of heaven. They all quickly advanced to the core of the Source phenomenon and saw, felt, or knew Source for what it actually is. None of them projected human religious or spiritual appearances or beliefs upon Source though they may have used religious language for lack of any other option. All of them readily accepted that we souls are actually Source, not humans. These factors indicate that expanded afterlife experiencers do not need the comfort and support of projected humanlike appearances in the afterlife, probably because the experience is not foreign to them.

Nothing in the data explains why these particular souls so quickly transitioned from human habitual thinking to open-minded acceptance of eternal truths unless it was their familiarity with the afterlife and Source.

The data also support several startling conclusions that are inconsistent with traditional religious and spiritual belief systems.

God is Not a Physical Being

The most astounding finding in my research is that expanded afterlife experiencers disprove the bedrock human belief that God the Creator is a human man who lives "up there somewhere." Every one of my subjects experienced God as a living, emoting, hugely intelligent, unconditionally loving and accepting phenomenon of some type rather than a being. It is not male or female. It has no body or fixed form. It cannot be accurately identified from experience obtained during human life. There are no human words for it. But all agree that this Source phenomenon is the Creator of, and is composed of, all of everything in existence including us.

The evidence in this study suggests that threshold afterlife experiencers who perceive a physical deity are manifesting their own religious or spiritual beliefs as reality within their afterlife experience. This is not criticism. Manifesting is not the same as projecting one's beliefs onto an otherwise disparate situation. Projection can involve perceiving a mental image as though it were real. Manifesting is

Source's supernatural ability to create entirely real physical reality. The reports of seeing manifested physical environments and people in the afterlife are true. However, those scenes and humanlike beings are real only to the soul manifesting them and are not permanent fixtures of the afterlife.

We Are Not Our Bodies

My research also demonstrates that these human animal bodies we inhabit do not represent our true identities. Human bodies are left behind when the soul crosses over into the afterlife. We are what humans call souls. Souls return to the afterlife upon human death.

Research group members learned through firsthand experience that human life is not real. Our true identity is we are Source. Our life is that of Source itself. We souls are unique mental characters/personalities included within the collective nature of Source as parts of its own consciousness and self-awareness. We have the same traits, powers, and nature though on a much smaller scale. We souls are eternal.

Source Created Us to Experience the Physical World

Humans have many theories about the meaning of life, some of which seem to be confirmed by NDEs and threshold afterlife experience reports. Expanded afterlife experiencers learned Source's purpose directly from it. Source constructed individual personalities within its mind, much as a novelist creates book characters within his/her mind, that it then injects into manifested physical matter creatures, plants, and inanimate objects. The goal is to obtain the inside view and sensations of life within the universe it manifested. Source wants to know what it feels like to be something other than itself, to know what it feels like to be limited, and to be able to relate to others unlike itself. Our purpose as souls is simply to live and gather those experiences. Nothing more profound. Incarnation and reincarnation are the methods we use.

We Experience Supernatural Events in the Afterlife

Expanded afterlife experiencers lived supernatural events unfathomable in human life.

One hundred percent of my research subjects who answered the NDERF questionnaire responded that they received "Knowings" on everything there is to know about the universe. Knowings differ from receiving information in the sense that Knowings impart not only data about a topic but also the sensation that one has personally lived that subject matter and gained knowledge of it through firsthand experience. As an analogy, think of the difference between receiving information and Knowing as the difference between getting a Microsoft download while on the outside of the database that holds it, and being inside the mind that has had the experiences and contains the database. Expanded afterlife experiencers get inside Source's mind.

During expanded life reviews, subjects were able to merge into other souls' minds and perceive events involving the subject from the other souls' perspectives, while simultaneously hearing their thoughts and feeling their emotions in response to the subject and the events.

Several subjects were able to review the past and/or preview the future while in the afterlife. Some witnessed Creation of the universe or our solar system.

Some subjects remembered having lived other lifetimes as other creatures and things, proving incarnation. Others watched souls moving in and out of Source's core identity to outer layers of its corona as they transposed to and from the physical universe via incarnation.

A few subjects enjoyed a profoundly different way to experience physical life through what I call virtual living. They were able to literally merge themselves into and with one or more other souls and witness and relive those other souls' physical lifetimes. The phenomenon is more expansive than what happens during a life review because the lives sampled do not include the subject, or possibly even human life. This stage of the afterlife appears to be an alternative to physical incarnation.

A few group members had missions on Earth that were monitored by councils of Light Beings.

Discussion

NDE researchers have proven that our consciousness, personality, and self-awareness can not only exist outside our body but also survive the body's death.

What then?

Those same scientific researchers study only the tip of the iceberg of what becomes of us souls once our physical hosts die. What happens to us later in the afterlife? After we absorb unconditional love and acceptance to our heart's content. After the life review. Fortunately, the evidence is available to us through the written or recorded accounts of those of us who have experienced more in the afterlife and have returned with firsthand, eyewitness testimony. Our expanded afterlife experience reports establish that we souls live forever as Source. My research subjects have explained how and why this is true, and what happens to us after we enter the afterlife.

Traditional academic NDE research is heavily grounded in medicine and science, both physical matter conventions. Yet, some researchers have extended their work beyond the physical world in order to truly understand the spiritual one. As Dr. Bruce Greyson says in *After*: "NDE research actually shows that by applying the methods of science to the nonphysical aspects of our world, we can describe reality much more accurately than if we limit our science to nothing but physical matter and energy."[296]

My research project goes further and accurately demonstrates the true nature of Source, the soul, and spiritual existence.

We expanded afterlife experiencers learned during eternal life phases that come after those described in NDE research that all souls eventually transition from thinking and believing like a human to thinking and Knowing like the eternal spiritual entity we are. The experiences that occur during the early phases after crossing over are largely manifestations of human life experiences carried over into a decidedly non-human state of being. Many souls do not realize they have left human life behind and attempt to recreate it in the threshold phase of the afterlife. This results in the many different descriptions of the physical scenery of heaven. Each soul that has a threshold afterlife experience manifests its own idea of what heaven should look like.

NDE research, on the whole, is a spectacular success because it proves beyond a shadow of a doubt that consciousness does not need a human body to survive. My research project leaps beyond the established science to analyze what medical NDE researchers perceive to be "outliers"—the reports of those afterlife experiences that go beyond the early, familiar phases or that include Knowings foreign to accepted religious and spiritual beliefs. The reported threshold afterlife phases are real, including leaving the body, sometimes going through blackness or a tunnel-like sensation, entering the afterlife, receiving Knowings, and having a life review. But the meaning given to afterlife manifestations and the knowledge souls receive in heaven in traditional medical NDE research is entirely human. It is normal to project one's understanding of life gained solely through human experience upon life, death, the afterlife, and God. Humans have no other frame of reference. But earthly experience does nothing to provide the truth about Source or eternal life, which is so far outside of human experience that some would call it alien. We expanded afterlife experiencers offer you the objective data we collected during our own most intimate communion with the Creator, the Source of all that exists—data that proves that we souls are eternal parts of the God humans traditionally worship.

22

My Expanded Afterlife and Near-death Experiences

I RELATE THE FULL DETAILS OF MY AFTERLIFE EXPERiences and everything I learned in the afterlife in my five previous books.[297] What follows is a summary.

Afterlife Experiences

First Death

I first entered the afterlife on March 14, 1994, with a combined attitude of scientific explorer and legal jurist sifting evidence. So, it is not surprising that my first afterlife experience focused almost exclusively on gaining answers to my spiritual questions. My spiritual journey began with a diagnosis of probable breast cancer necessitating an excisional biopsy (more extensive than a lumpectomy) for confirmation.

Before I died the first time, I had been a devoted spiritual truth seeker. I was reared in the Roman Catholic Church and attended Catholic schools for twelve years. I attended a Methodist college for four years full-time, and two years part-time, while earning a Bachelor of Science degree with a double major in Chemistry and Biology (Anatomy and Physiology concentration), and a Bachelor of Arts degree in Psychology. I thereafter earned my Doctorate in Law. I practiced law full-time for thirty-six years, starting out as an employment lawyer and litigator, but switching to health law before it really started booming as a specialty. Half of my practice years were spent as an associate and then partner in one of the largest and most prestigious law firms in my state. The other half I spent as a solo practitioner. The dividing line between practice settings was my body's death in 1994. That changed everything.

Before my afterlife journey began, I was healthy, wide-awake, under the influence of no medication except a local anesthetic to the skin of my breast, had nothing to eat or drink for more than twelve hours, and had no idea my body was dying. Consequently, I had no reason to mentally concoct scenarios about what dying would be like. In fact, I believe I was chosen to have this experience precisely *because* I was healthy and in the unique medical and personal position to question what was happening to me, analyzing it like a scientist and jurist. I did not lose one moment of awareness during my body's dying and death, or during my own crossing over and afterlife. I was conscious throughout and remember it all in much more detail than given here.

Briefly, I had a pre-cancer surgery invasive procedure where a radiologist twice stuck a large needle with a wire inside it into my breast to mark the tumors for the surgeon. As a result, my body died from a combination of anaphylactic shock from the anesthetic, very low blood sugar, and tachycardia. I went into the Light, was completely saturated with unconditional love, bliss, and acceptance; saw my body in the mammography room while looking down from heaven; manifested (created) several extremely real earthlike environments just by thinking the words; met my five most beloved friends from eternal life (not from incarnated life), who appeared in the form of Light Beings; had an in-depth life review and remembered hundreds or thousands of other physical lifetimes; and was given the answers to my most burning questions: What is God? What am I? What does God expect of me? What's the purpose of life? Where's heaven? Where's hell? and What's the one true religion? I asked this last question because the Catholic Church claimed to be the one true religion. The answers downloaded into my mind were nothing remotely similar to what the Catholic Church had taught me. I was angry that I had not been given the correct answers before I died, as I assumed everyone else on Earth knew this information but me. In response, I was shown a documentary-type Earth history, focused on how religions developed, to explain why my religion had taught me erroneous information about life, death, God, and the afterlife. After that I moved into the next stage of afterlife, where I literally merged my Energy and beingness into my five Light Being friends so that I could experience their physical lifetimes similar to a virtual reality game, only much more sophisticated. The five of us then moved through the Energy field that I call Source and others call God, Yahweh, Allah, the Creator, etc., to its core. While deeper inside Source's Energy field, I watched Creation of our universe as the meaning and purpose of life was explained to me by Source itself.

I was told in no uncertain terms that you and I are actually Source, simply playing a role similar to how we currently play roles in our dreams. I was told that I had never been separated from Source and was in fact an integrated part of its consciousness and self-awareness that had merely temporarily inhabited a human animal. The purpose of life is to allow Source to experience the feelings and sensations of the universe it created comparable to how we experience our dreams through the dream-character version of ourselves. Source imposed amnesia on the parts of itself that incarnate into physical matter so that it could let its imagination run wild without restraints imposed by its own unconditionally loving nature. Therefore, when our bodies die, we simply wake up from the dream of human life and resume living spiritual life in what we call the afterlife.

At the end of this extensive afterlife adventure, I accepted the mission to experience unconditional love during Nanci's life and to tell anyone who would listen what I had learned in the afterlife. Now, having read or heard nearly two thousand near-death and afterlife experience accounts, it continues to surprise me that I remember so much more than others of what happened to me and what I learned in the afterlife.

I personally experienced all of the phases of the afterlife described in NDE research literature. In addition, I experienced much, much more after my life review, which is the last phase generally reported by threshold afterlife experiencers. I have come to believe that I processed through so many afterlife phases because it took that many different experiences to convince me that my former Catholic and spiritual beliefs were erroneous. I did not want to give them up. I resisted accepting the Knowings I received even though I was certain in my heart-of-hearts that they are true. I *remembered* knowing them before incarnating into Nanci. I continued to resist the Knowings for seven years after I returned to Nanci's life. While my need for proof may have made me a recalcitrant visitor to heaven, perhaps it makes me the perfect reporter of how and why consciousness survives human death and what awaits us in the rest of eternal life.

Fortunately, the tumors removed from my breast were pre-cancerous and not malignant. I flew to Baltimore, Maryland, the next day and gave a presentation to a packed room of healthcare lawyers from around the country.

Second Death - First Council Meeting

Many death survivors make dramatic changes in their lifestyles. I did not—at least not immediately. Consequently, the Council of Light Beings monitoring my mission called me back to the afterlife within months after my initial visit. I have no idea why I died this time. I was alone in my office when I suddenly recognized the afterlife before my eyes. Several brilliant Beings unknown to me seemed to be gathered in a type of formal conclave, similar to a legal hearing or tribunal, with my new mission on Earth the focus of attention. It was evident to me that each person in the group had a keen interest in how well I was performing it, and that they were not even slightly satisfied. All I can remember now is that the council felt I was not pursuing what I had planned while inside Source. They were there to guide and support me in meeting my goal. I do not remember exactly what they expected me to do, but my conscious mind later interpreted my instructions to be to start pursuing my new spiritual mission.

Third Death – Second Council Meeting

A second meeting with my council seven years later was more dramatic and traumatic, probably because my body was very ill. In 2001, I once again found myself in the afterlife at what appeared to be a hastily called gathering in the Light. The vague, semi-transparent earthlike setting appeared more like a conference room than a formal hearing chamber. The Light Being members were settling into chairs around a crowded table when I arrived. Three council members joined the group after I did, lending credence to my impression of spontaneity.

Two of the last three Beings to enter displayed the faces of Nanci's biological parents for only a second before that illusion was replaced with their luminescent Light Being forms. My Earthly parents' participation both thrilled and shocked me. Their human hosts had died years before this council meeting, so I was excited to see their faces again even for a moment. I felt as though the souls who had played my human parents had given me the wonderful gift of recognition to reassure me in advance of the meeting about to take place. But having my human parents on my council disturbed me, as I felt I had not respected their guidance as well as I should have while they were on earth. (I might have taken them more seriously had I known they were going to be on my council!)

The last council member to arrive appeared in human form, projected a sense of rush, and remained in human appearance for some time. When I raised my gaze to his, he looked me straight in the eye and mouthed the word "surprise!" And I *was* surprised! I recognized this Being as a human I know this life as Jeff. An alarm went off in my mind seeing someone from my human life, as if some emergency had triggered a recall of even undercover agents.

This second council meeting was brief and to the point. The

Light Beings told me that extremely difficult times lay ahead—times that would cause me grave suffering. The council acknowledged my body's illness and weakened condition, intimating my human might not have sufficient strength to revive. The council gave me the option of staying in human life, or reawakening to spiritual level right then to avoid the suffering. They made it clear that if I chose to leave human life, it would not be considered a breach of contract on my part. (The mission to tell everyone what I learned in the afterlife was the contract they were relieving me from performing.) If I stayed in my human host, they said, I will endure much suffering. Despite what they told me about my future, I chose to stay in human form. I did not want to be a failure at the mission Source gave me, I felt a duty to inform my fellow incarnated souls of what I experienced, and I wanted to see what was going to happen in Earth's future.

From the time I first visited the afterlife in 1994, I have had what I call "pop-in" afterlife experiences. I had one in 1999 during deep meditation. During this pop-in, I had another life review that was not only current to that date but also included my future. The detail was phenomenal. I experienced my own thoughts and actions as well as those of the people with whom I was interacting. For example, I watched a meeting I attended with several attorneys in my old law firm. The young attorney I was mentoring sat across the table from me. During the life review, I saw myself kick his leg under the table. He felt that I was rebuking him for something he had said. I was actually just clumsily crossing my legs. During this life review, I understood that our intentions are as important as our actions.

I had another pop-in to the afterlife while writing my book *BACKWARDS Beliefs: Revealing Eternal Truths Hidden in Religions*. I sat typing at my kitchen table when all of a sudden I was back in the afterlife getting a download of Knowings. This download included every single thing that had ever been or ever will be written by every human throughout eternity, including phone call slips, homework papers, doodles, Post-it notes, music, literature, and everything else. I read and understood it all at the same time. Certain passages were highlighted in gold light, so I concentrated on them. These records showed me that there has been an underlying theme of "we are all one" running throughout human writings, clearly emanating from the memories of souls within. The only example I can remember now is a song by the Canadian group named Nickelback.

My Near-Death/Out-of-Body Experiences

Dying is not good for your health. My 1994 experience, which lasted about a half hour of human time, left me with brain damage from lack of oxygen and with several disabilities. At one point, I developed an inability to keep my body temperature, blood pressure, and heart rate at normal levels. My cardiologist prescribed various medications over the years in a failed attempt to normalize my vital signs. One of them put me in the hospital in 2001.

First Hospital NDE

On August 25, 2001, I began having extreme chills, nausea, and what felt to me like convulsions. I tried to sleep it off. I got up once to go to the kitchen to get some water. On the way back to my bedroom, I projectile-vomited a large quantity of acidic fluid all over the hardwood floor in the dining room. Afterward, (and some of you ladies will get this) I mopped up the vomit and changed clothes. *Then* I called 911.

By the time I got to the Emergency Department, I could barely speak. I told the ER physician that I felt like I was having convulsions, could not think clearly, and could barely use my mouth to speak. The inability to speak felt exactly like it did when I came back into my body after my 1994 afterlife experience. It was as though I was speaking through a long tube that connected to my body's mouth and I didn't have much control over what went through the tube. I recognized that I was getting out of body. The doctor took blood for labs and sent me for an MRI.

While I was lying on a gurney in the hallway outside the MRI room, I sat up out of my body. I was still encased in the body's legs. But from the waist up, I was out-of-body—sitting up looking down the hallway and through walls. I saw my ER doctor through the walls forming a corner of the hallway before he rounded the corner. He was running with an IV bag in his hands. He panted that he knew what was wrong with me. My lab results showed an extremely low blood sodium level. He said a few points lower, and I would have died. I heard this with my spiritual ears and laid back down into my body. The doctor hooked up the IV bag of sodium right there in the hallway and admitted me to the hospital. It took three days for my blood sodium levels to normalize. It took a couple of years to recover from the erratic vital signs fluctuations, and months to regain my strength.

Second Hospital NDE

On June 1, 2011, after being diagnosed with aggressive metastatic breast cancer, I was in the local cancer hospital for a lumpectomy and sentinel lymph node biopsy. My anesthesiologist was intrigued when she found out I had had NDEs and spent 45 minutes talking with me before the surgery started. At the beginning of the surgery, she gave me a lighter level of anesthesia than general anesthesia and that has fewer side effects. I was immediately unconscious.

A few minutes later, I was conscious! But I was conscious of being outside my body. I was wandering around the surgical suite outside of the OR where my body was on the surgical table. I looking around at everything. At one point, I looked through either a window or a wall (I can see through walls when I'm out of body) and saw a body on a table with two surgeons, an anesthesiologist, and one or more nurses around it. I recognized it as surgery from watching way too many episodes of *Grey's Anatomy* on TV. Suddenly, I realized that I was the one who was supposed to be on that operating table. I said to myself, "I'm supposed to be in there"—meaning in that body. Immediately upon having that thought I was back in the body and unconscious again.

The surgery was a short one. My surgeon told me in the recovery room that the surgical prep took 5 minutes, the surgery took 10 minutes, but it took the nurse 20 minutes to read out loud the extensive list of my medical allergies before they could get started. I had been seeing this doctor since my 1994 breast cancer surgery and this flippant comment about my allergies is the first funny thing I had ever heard him say.

Third Hospital NDE

On July 6, 2011, I had my third and fourth breast cancer surgeries for the year. My surgical oncologist performed a radical mastectomy and lymph node dissection, following the same needle localization procedure during which I died in 1994. Because of the failed lumpectomy in June, I told my surgeon he had my permission to remove anything he needed to on the outside of my ribcage, but to leave intact everything inside my ribcage. He took me at my word. The mastectomy was an old-fashioned one consisting of removing everything, including breast tissue, muscles, blood vessels, lymph nodes and lymphatic vessels, on my left ribcage from breastbone to shoulder blade in my back. Everything but the skin over my ribs was removed. The surgeon left me one small piece of pectoralis muscle up by my shoulder, sewn down to the ribcage, so that my left arm would not flop around. The lymph node dissection removed all the lymph nodes, lymph vessels, fat, fascia and much of the skin in my left armpit.

The next day, the anesthesiologist came to my hospital room to tell me what had happened during the surgery. She said that my heart rate and blood pressure dropped dramatically, and she had to resuscitate me. She said she knew I had a "do not resuscitate" order on my chart but could not bear to let me go after reading my first book.

I told her that I saw part of the surgery. I remember hovering over the operating table and seeing three people bellied up to it. My surgeon was on my right. His resident was on my left. And the anesthesiologist was at the top of my head. I heard someone say, "Wow, that's a beautiful incision." And it was – particularly since it had to be closed with glue because I'm allergic to sutures. Most mastectomies leave the woman with a big pooch under her arm. I have none. The anesthesiologist confirmed the location of the three doctors and said that she was the one who had made that exact comment about the incision.

Over the years, I have had several out-of-body experiences unrelated to any medical condition. Thus, I can compare the two types of OOB events.

My Shared/Empathic NDE

The phenomenon in which a second person in close physical proximity to, or emotional connection with a dying person sees, hears, or feels what the dying person experiences in the initial stages of crossing over is called an "empathic" or "shared" NDE.

My mother died permanently the same year I died temporarily for the first time. She was in a hospital in hospice care. I was privileged to sit beside her bed holding her hand. Several months before her death, she and I discussed what to expect upon dying, as we had both had NDEs. Presumably because I was holding her hand and concentrating so intensely, I could see that all she saw was blackness after leaving the body. I literally saw the blackness she saw. She knew to look for the Light but did not see it right away. I watched her look around for it. I felt her panic and suddenly jump back into the physical world outside her former body. I felt her invisible presence flit around the hospital room in a desperate attempt to find the Light. This was confirmed by the priest in the room, who jumped back at the very moment I felt mom go by him. He said "she's out" and I said yes. She eventually returned to the body and passed peacefully sometime later.

If ever you had any doubt that we souls remain conscious after the body's death, having an empathic NDE would convince you.

Notes

* Aaron M (2015), "NDE 8350," Near-Death Experience Research Foundation, https:// www.nderf.org/Experiences/1aaron_m_nde.html.

1. Wayne H. (2006-2007), "NDE 2543," Near-Death Experience Research Foundation, https://www.nderf.org/ Exp.riences/1wayne_h_nde.html.

2. Sue C. (2005), "NDE 6159," Near-Death Experience Research Foundation, https://www.nderf.org/ Experiences/1sue_c_nde.html.

3. The host of the "Heaven Awaits You" YouTube channel reads Anonymous's account in a video posted under the title "I Died and Found out the Truth About God." That video is posted at https://www.youtube.com/watch?v-QNplf7WjGMc. I found the written account in the NDERF Explorer under the name ErinRae G. However, when you search her name in the NDERF database, the account that appears under ErinRae G.'s name is completely different than Anonymous's account.

4. Ibid.

5. Jennifer J. (2001), "NDE 7510," Near-Death Experience Research Foundation, https://www.nderf.org/Experiences/ 1jennifer_j_ndes.html.

6. Mira Sai (1994), "NDE 7602," Near-Death Experience Research Foundation, https://www.nderf.org/Experiences/ 1mira_s_nde.html.

7. Bridget F. (1995), "NDE 3648," Near-Death Experience Research Foundation, https://www.nderf.org/Experiences/ 1bridget_f_nde.html.

8. Valeska K. (2019), "FDE 8750," Near-Death Experience Research Foundation, https://www.nderf.org/Experiences/ 1valeska_k_nde.html.

9. Malinda K. (1991), "NDE 3362," Near-Death Experience Research Foundation, https://www.nderf.org/Experiences/1malinda_k_nde.html.

10. Nanci L. Danison, *BACKWARDS: Returning to Our Source for Answers,* (Columbus, OH: A.P. Lee & Co. 2007), 12-13.

11. Chantal L. (1991), "NDE 6428," Near-Death Experience Research Foundation, https://www.nderf.org/ Experiences/1chantal_l_nde.html.

12. Aaron M. (2015), "NDE 8350," Near-Death Experience Research Foundation, https://www.nderf.org/Experiences /1aaron_m_nde.html.

13. Joan L.H. (1985), "NDE 6896," Near-Death Experience Research Foundation, https://www.nderf.org/Experiences/ 1joan_lh_nde.html.

14. Ron Kruger (1962), "NDE 2408," Near-Death Experience Research Foundation, https://www.nderf.org/ Experiences/1ron_k_nde.html.

15. Kathryn H. (2009), "NDE 6975," Near-Death Experience Research Foundation, https://www.nderf.org/ Experiences/1kathryn_h_nde.html.

16. Ron Kruger, "NDE 2408."

17. D.W. (1984), "NDE 6106/3587," Near-Death Experience Research Foundation, https://www.nderf.org/Experiences/1dw_nde.html.

18. Ibid.

19. Demi B. (1962), "NDE 6405," Near-Death Experience Research Foundation, https://www .nderf.org/Experiences/demi_b_nde.html.

20. William W. (1985), "NDE 6292," Near-Death Experience Research Foundation, https://www.nderf.org/ Experiences/1william_w_nde.html.

21. Bobby R. (unknown date), "NDE 8010," Near-Death Experience Research Foundation, https://www.nderf.org/ Experiences/1bobby_r_nde.html.

22. William H. (2003), "NDE 7340," Near-Death Experience Research Foundation, https://www.nderf.org/ Experiences/1william_h_nde.html.

23. Wayne H. (2006-7), "NDE 2543," Near-Death Experience Research Foundation, https://www.nderf.org/Experiences/1wayne_h_nde.html.

24. Sue C. (2005), "NDE 6159."

25. Anonymous account in NDERF Explorer.

26. Jennifer J. (2001), "NDE 7510."

27. Mira Sai (1994), "NDE 7602."

28. Bridget F. (1995), "NDE 3648."

29. Valeska K. (2019), "FDE 8750."

30. Malinda K. (1991), "NDE 3362."

31. Chantal L. (1991), "NDE 6428."

32. Aaron M. (2015), "NDE 8350."

33. Ibid.

34. Joan L.H. (1985), "NDE 6896."

35. Ron Kruger (1962), "NDE 2408."

36. Ibid.

37. Kathryn H. (2009), "NDE 6975."

38. D.W. (1984), "NDE 6106," Answer to Ques. 49.

39. Demi B. (1962), "NDE 6405."

40. William H. (2003), "NDE 7340."

41. Andy P. (Petro) (2005), "NDE 2335," Near-Death Experience Research Foundation, https://www.nderf.org/ Experiences/1andrew_p_nde.html.

42. Bobby R. (unknown date), "NDE 8010."

43. William H. (2003), "NDE 7340."

44. Yazmine S. (1978), "NDE 6992," Near-Death Experience Research Foundation, https://www.nderf.org/ Experiences/1yazmine_s_nde.html.

45. Robyn (2006), "NDE 6636," Near-Death Experience Research Foundation, https://www.nderf.org/ Experiences/1robyn_nde.html.

46. You can hear Andy tell his story at http://ndestories.org/andy-petro. He also has a video on the "Let's Talk Near-death" Channel on YouTube.

47. Doug F. (1979), "NDE 8422," Near-Death Experience Research Foundation, https://www.nderf.org/ Experiences/1doug_f_nde.html.

48. Emanuele (2018), "NDE 8722," Near-Death Experience Research Foundation, https://www.nderf.org/ Experiences/1emanuele_nde.html.

49. Kathryn H. (2009), "NDE 6975," Answer to Ques. 41.

50. Wayne H. (2006-2007), "NDE 2543."

51. Natalie Sudman, *Application of Impossible Things: My Near-Death Experience in Iraq* (Huntsville AR: Ozark Mt. Publishing 2012), 28, quoted with permission of Ozark Mountain Publishing.

52. Aaron M. (2015), "NDE 8350."

53. Nanci L. Danison, *BACKWARDS: Returning to Our Source for Answers* (Columbus, OH: A.P. Lee & Co. 2007), 12-13.

54. Valeska K. (2019), "FDE 8750."

55. Henry W. (2000), "NDE 3624."

56. D.W. (1984), "NDE 6106."

57. Ibid.

58. Ibid.

59. Aaron M. (2015), "NDE 8350."

60. Ibid.

61. Andy Petro and Kirsty Salisbury, "The NDE of Andy Petro," Let's Talk Near-death, YouTube, https://www.youtube.com/watch?v=vIb8kArzjLk&t=829s at 13.55 to 14:08 and at 14:08 to 14:40.

62. Ibid.

63. Andy Petro, "The NDE of Andy Petro," at 18:00 to 19:29.

64. Andy Petro (2005), "NDE 16071/2335," Near-death Research Experience Foundation, https://www.nderf.org/ Experiences/1andrew_p_nde.html.

65. AAnke Evertz on Thanatos TV on YouTube, https://www.youtube.com/watch?v=O2whJPweTkQ at 32:11 to 32:44.

66. Leonard (2009), "NDE 4046," Near-Death Experience Research Foundation, https://www.nderf.org/Experiences/1leondard_nde.html.

67. Chantal L. (1991), "NDE 6428."

68. D.W. (1984), "NDE 6106," Answer to Ques. 47.

69. D.W. (1984), "NDE 6106."

70. Ibid.

71. Dea M. (2015), "NDE 4281."

72. Emanuele (2018), "NDE 8722."

73. Ibid.

74. Joan L.H. (1985), "NDE 6896."

75. Ibid.

76. Natalie Sudman, *Application of Impossible Things,* 74.

77. Kathryn H. (2009), "NDE 6975."

78. William H. (2003), "NDE 7340."

79. William W. (1985), "NDE 6292".

80. Yazmine S. (1978), "NDE 6992."

81. Demi B. (1962), "NDE 6405."

82. Ibid. at Answer to Ques. 47.

83. Gwen J. (2006), "NDE 6429," Near-Death Experience Research Foundation, https://www.nderf.org/Experiences/ 1gwen_j_nde.html, at Answer to Ques. 27.

84. Emanuele (2018), "NDE 8722."

85. Jennifer J. (2001), "NDE 7510."

86. Ibid. at Answer to Ques. 47 and 49.

87. Malinda K. (1991), "NDE 3362."

88. Ibid.

89. Mira Sai (1994), "NDE 7602."

90. Ibid.

91. Pamela K. (1993), "NDE 4649," Near-death Experience Research Foundation, https://www.nderf.org/Experiences/ 1pamela_k_nde.html.

92. Robyn (2006), "NDE 6636."

93. Peter N. (1974), "NDEs 6584/10105."

94. Barbara R. Rommer, MD, *Blessing in Disguise: Another Side of the Near-death Experience* (St. Paul, MN: Llewellyn Publications 2000), 137.

95. William H. (2003), "NDE 7340."

96. Yazmine S. (1978), "NDE 6992."

97. Malinda K. (1991), "NDE 3362", Answer to Ques. 36.

98. Leonard (2009), "NDE 4046."

99. Chantal L. (1991), "NDE 6428."

100. Ibid., in a response to a letter from Dr. Jeffrey Long.

101. Emanuele (2018), "NDE 8722."

102. All the details of Creation that I remember are included in the chapter titled "Creation of the Universe" in my book *BACKWARDS Beliefs: Revealing Eternal Truths Hidden in Religions* (Columbus, OH: A.P. Lee & Co., Ltd. 2011).

103. We experience something like expelling Thought Energy when we have to concentrate and work hard to solve an important problem, or take a life-changing test, or disarm a bomb, for example. Tasks like these can leave us feeling drained of energy.

104. Ron Kruger (1962), "NDE 2408."

105. Aaron M. (2015), "NDE 8350."

106. Leonard (2009), "NDE 4046", Answer to Ques. 41.

107. The full details I recall from watching creation of souls in the afterlife is contained in my book *BACKWARDS Beliefs: Revealing Eternal Truths Hidden in Religions* (Columbus, OH: A.P. Lee & Co. 2011).

108. Gwen J. (2006), "NDE 6429", Answer to Ques. 43.

109. Ibid.

110. Sue C. (2005), "NDE 6159."

111. Emanuele (2018), "NDE 8722."

112. Peter N. (1974), "NDEs 6584/10105."

113. D.W. (1984), "NDE 6106."

114. Ibid.

115. Henry W. (2000), "NDE 3624."

116. Leonard (2009), "NDE 4046." Answer to Ques. 41.

117. Aaron M. (2015), "NDE 8350."

118. Robyn (2006), "NDE 6636."

119. Sue C. (2005), "NDE 6159," Answer to Ques. 52.

120. Malinda K. (1991), "NDE 3362."

121. Andy Petro on "Let's Talk Near-death" at roughly 14:20.

122. Chantal L. (1991), "NDE 6428."

123. Robyn (2006), "NDE 6636."

124. Leonard (2009), "NDE 4046."

125. Aaron M. (2015), "NDE 8350."

126. Ibid.

127. Chantal L., "NDE 6428," in response to a letter from Dr. Jeffrey Long.

128. Anonymous, "I Died and Found out the Truth About God," posted at https://www.youtube.com/watch?v-QNplf7WjGMc at 14:54 – 15:05.

129. Joan L.H., "NDE 6896," Answer to Ques. 46.

130. Linda G. (1999), "NDE 3649," Answer to Ques. 38. I totally understand what Linda means because when I was in the afterlife I dearly missed chocolate. I might have considered coming back just for that.

131. Doug F. (1979), "NDE 8422."

132. Natalie S. (Sudman) (2007), "NDE 6246," Answer to Ques. 50, Near-Death Experience Research Foundation, https://www.nderf.org/Experiences/ 1natalie_s_ html.

133. Henry W. (2000), "NDE 3624."

134. Jennifer J. (2001), "NDE 7510," Answer. to Ques. 38.

135. Ibid.

136. Valeska K. (2019), "FDE 8750."

137. Kenneth Ring, PhD. 1982. *Life at Death: A Scientific Investigation of the Near-Death Experience* (New York: Quill 1982), 52. (Orig. pub. 1980.)

138. Aaron M. (2015), "NDE 8350."

139. Jennifer J. (2001), "NDE 7510."

140. Ibid.

141. D.W. (1984), "NDE 6106."

142. Mira S. (Sai) (1994), "NDE 7602."

143. Wayne H. (2006-2007), "NDE 2543."

144. Ibid.

145. Leonard (2009), "NDE 4046."

146. Ron K. (Kruger) (1962), "NDE 2408."

147. Henry W. (2000), "NDE 3624."

148. I was a practicing health lawyer at the time and had reviewed thousands of medical records for correct coding and billing in accordance with CPT coding guidelines. CPT stands for "Current Procedural Terminology" and is published by the American Medical Association.

149. Sue C. (2005), "NDE 6159."

150. Beatrice W. (1996), "NDE 9251," Near-Death Experience Research Foundation, https://www.nderf.org/Experiences/1beatrice_w_nde.html.

151. Andy P. (Petro) (2005), "NDE 2335."

152. D.W. (1984), "NDE 6101."

153. Joan L.H. (1985), "NDE 6896."

154. Anonymous (2021), "I Died and Found out the Truth About God" video.

155. Ibid.

156. Dylan Donnelly, "Life after death: NDE survivor revisits 'vivid and crisp' scene from childhood," *Express,* March 11, 2021, https://www.express.co.uk/news/weird/1408337/Life-after-death-Near-death-experience-bruce-Greyson-Gregg-nome-drowning-visions-ont?fbclid IwAR3wodLk4zKxGRPhAUvlib25Q1Wp-m2OoybfyrN9zzkPdQnjnCXE6JQHOs04.

157. Personal conversation.

158. Leonard (2009), "NDE 4046."

159. Anke Evertz on Thanatos TV, https://www.youtube.com/watch?v=O2whJPweTkQ at 15:00.

160. Carol I. (1980), "NDE 5188," Near-Death Experience Research Foundation, https://www.nderf.org/Experiences/ carol_i_nde.html.

161. Bobby R. (unknown date), "NDE 8010."

162. Anke Evertz video at 15:00 to 18:00.

163. Leonard (2009), "NDE 4046."

164. Andrew P. (Petro) (1955), "NDE 2335."

165. D.W. (1984), "NDE 6106."

166. Pamela K. (1993), "NDE 4649."

167. Peter N. (1974), "NDEs 6584/10105."

168. Ibid.

169. Ron K. (Kruger) (1962), "NDE 2408."

170. William H. (2003), "NDE 7340."

171. Doug F. (1979), "NDE 8422."

172. Andy Petro, *Let's Talk Near-Death – The NDE of Andy Petro,* https://www.youtube.com/watch?v=vlb8kltrzjLK.

173. Ron K. (Kruger) (1962), "NDE 2408."

174. Wayne H. (2006-7), "NDE 2543."

175. Ibid.

176. Emanuele (2018), "NDE 8722."

177. Mira S. (Sai) (1994), "NDE 7602."

178. Ibid.

179. Pamela K. (1993), "NDE 4649."

180. Sudman, *Application of Impossible Things,* 53.

181. Jeffrey Olsen, *I Knew Their Hearts* (Springville, UT: Plain Sight 2012), 32.

182. Kane D. (2002), NDE 3692, Near-Death Experience Research Foundation, www.nderf.org/Experiences/ 1kane_d_nde.html.

183. Aaron M. (2015), "NDE 8350."

184. My observation has been that all souls that escape the body, however briefly, and regardless of whether they enter the afterlife, reasonably truly and deeply believe what their out-of-body (OOB) thoughts revealed to them. Knowings I received in the afterlife showed me that the difference between OOB insights and receiving Knowings is the content. Knowings convey eternal truths previously unsuspected by the soul receiving them. They have little or no similarity to the soul's religious or spiritual beliefs. OOB thoughts, in contrast, generally repeat and reaffirm information to which the soul has been previously exposed during the body's lifetime but the body has forgotten. Retrieving these thoughts is possible because, once out of body, souls have total recall of everything to which they have ever been exposed, including information the body itself cannot remember or to which he/she paid little attention at the time. This data includes all of the religious and spiritual precepts learned as a small child. It also includes all fiction the body has been exposed to, including TV, movies, books, games, dreams, and more. OOB spiritual revelations generally integrate everything the soul has ever read, heard, thought or imagined during human life, all tied up in a very convincing stream of thought. That information is, however, a mixture of both fact and fiction. It includes human thinking, speculation, mythology, and superstition with perhaps a thin sprinkling of actual Knowings in the mix.

185. Joan L.H. (1985), "NDE 6896."

186. D.W. (1984), "NDE 6106."

187. Anonymous (2021), "I Died and Found out the Truth About God" video at https://www.youtube.com/watch?v-QNplf7WjGMc.

188. Peter N. (1974), "NDE 6584."

189. Ibid.

190. Ibid.

191. Wayne H. (2006-7), "NDE 2543."

192. Henry W. (2000), "NDE 3624."

193. Yazmine S. (1978), "NDE 6992."

194. William W. (1985), "NDE 6292."

195. Burke (1988), "NDE 78," Near-Death Experience Research Foundation, www.nderf.org/Experiences/1burke nde.html.

196. Jeffrey Long, MD with Paul Perry, *God and the Afterlife* (New York: HarperCollins Publishers), 18, paraphrasing the report of, Robyn (2006), "NDE 6636." Quoted with permission of HarperCollins Publishers.

197. D.W. (1984), "NDE 6106."

198. Anonymous (2021), "I Died and Found out the Truth About God," at around 13:00.

199. Ibid., at around 13:57.

200. Chantal L. (1991), "NDE 6428."

201. Ibid., response to a letter from Dr. Jeffrey Long, the NDERF founder.

202. Jennifer J. (2001), "NDE 7510."

203. Leonard (2009), "NDE 4046."

204. My previous five books detail everything I can remember about Knowings on various topics. Additional explanations of some Knowings appear in my Zoom videos, which are available at https://nancidanison.com, and in YouTube videos at

Nanci Danison Channel.

205. The detailed answers to these questions constitute the first 90 pages of my book *BACKWARDS: Returning to Our Source for Answers,* as well as chapters in the *BACKWARDS Guidebook,* and *Answers From The Afterlife.*

206. Dannion Brinkley with Paul Perry, *Saved by the Light: the True Story of a Man Who Died Twice and the Profound Revelations He Received* (New York: HarperCollins Publishers 1994), 57.

207. Joan L.H. (1985), "NDE 6896."

208. My book *BACKWARDS: Revealing Eternal Truths Hidden in Religions* details what I remember from watching a documentary-type vision of Earth's history and future with an emphasis on how religions developed over the millennia.

209. Jean R. (1981), "NDE 6166," Near-Death Experience Research Foundation, https://www. nderf.org/Experiences/1jean_r_nde_6166.html.

210. Jennifer J. (2001), "NDE 7510."

211. Anonymous (2021), "I Died and Found out the Truth About God," at around 13:27.

212. Ibid., at around 13:47.

213. Joan L.H. (1985), "NDE 6896," Answer to Ques. 25.

214. Those questions are: What is God? What am I? What's the purpose of life? Why am I on Earth? Where's heaven? Where's hell? and What's the one true religion?

215. This vision is described in detail in my book *BACKWARDS Beliefs: Revealing Eternal Truths Hidden in Religions* (Columbus, OH: A.P. Lee & Co., Ltd. 2011).

216. Ron K. (Kruger) (1962), "NDE 2408."

217. Aaron M. (2015), "NDE 8350."

218. Henry W. (2000), "NDE 3624."

219. Ron K. (Kruger) (1962), "NDE 2408."

220. Peter N. (1974), "NDE 6584."

221. William H. (Horden) (2003), "NDE 7340.

222. Chantal L. (1991), "NDE 6428."

223. Sudman, *Application of Impossible Things,* 3-4.

224. Robyn (2006), "NDE 6636."

225. Sudman, 15.

226. The NDERF questionnaire includes a section where the experiencer can check boxes indicating various elements that were present in their NDE. Apparently, in the past, the questionnaire gave the option of checking a box labeled "a landscape or city" as an element. Chantal L, Doug F and Leonard checked this box but there is nothing in their narrative accounts or other answers that would seem to support that they did see a landscape or city in the afterlife. They also checked the box for "clearly mystical or unearthly realm." Chantal commented that all of her experience occurred in this other world and other state. Doug's description of the unearthly realm is merely that it was less dense, more expanded, and filled with light. Leonard's description (which I roughly translated myself from French) was an environment full of light in colors not found on Earth. These comments would seem to contradict the checkmark for a physical landscape or city as that phrase is used in medical NDE core elements descriptions.

227. Everything I remember learning in the afterlife about manifesting is detailed in my book *Create a New Reality: Move Beyond Law of Attraction Theory* (Columbus, OH: A.P. Lee & Co., Ltd. 2018).

228. Mira S. (Sai) (1994), "NDE 7602."

229. Chantal L. (1991), "NDE 6428."

230. Emanuele (2018), "NDE 8722."

231. Dea M. (2008), "NDE 4281."

232. Leonard (2009), "NDE 4046."

233. Rommer, *Blessing in Disguise,* 137.

234. Aaron M. (2015), "NDE 8350."

235. Andy P. (Petro) (2005), "NDE 2335."

236. Doug F. (1979), "NDE 8422."

237. Peter N. (1974), "NDEs 6584/10105."

238. Ron K. (Kruger) (1962), "NDE 724."

239. William H. (Horden) (2003), "NDE 7340." William Horden calls his body his lower soul and himself as soul his higher soul.

240. Bobby R. (unknown date), "NDE 8010."

241. The exact wording of the three answers may have changed a little over the years, but the concepts remained the same. I have read slightly different versions of the question in my subjects' written accounts. I also wrote down the multiple-choice answers used in the current questionnaire. But I have no access to previous versions of the answers that were offered to submitters at various times in the past.

242. Consequently, it is possible that researchers who used the NDERF database, and who relied solely upon the answers to this question without reading the full narratives, have overreported the number of barrier or boundary incidents in NDE accounts.

243. D.W. (1984), "NDE 6106."

244. Aaron M. (2015), "NDE 8350."

245. Ibid.

246. Ibid.

247. Bridget F. (1995), "NDE 3648." Although Bridget answers the questionnaire about reaching a boundary in the affirmative, she explains in later answers that she was referring to realizing that her human body would die if she did not return to it.

248. Sue C. (2005), "NDE 6159."

249. Leonard (2009), "NDE 4046."

250. Dea M. (2008), "NDE 4281."

251. Yazmine S. (1978), "NDE 6992."

252. Andy P. (Petro) (2005), "NDE 2335," Answer to Ques. 27, 29.

253. Chantal L. (1991), "NDE 6428," Answer to Ques. 36.

254. D.W. (1984), "NDE 6101," Answer to Ques. 40.

255. Joan L.H. (1985), "NDE 6896," Answer to Ques. 34.

256. Jennifer J. (2001), "NDE 7510," Answer to Ques. 35.

257. Bobby R. (unknown date), "NDE 8010," Answer to Ques. 36.

258. Demi B. (1962), "NDE 6405," Answer to Ques. 38.

259. Malinda K. (1991), "NDE 3362," Answer to Ques. 26.

260. William W. (1985), "NDE 6292," Answer to Ques. 33.

261. Peter N. (1974), "NDEs 6584/10105," Answer to Ques. 35.

262. Robyn (2006), "NDE 6636," Answer to Ques. 35.

263. Aaron M. (2015), "NDE 8350," Answer to Ques. 44.

264. Sue C. (2005), "NDE 6159," Answer to Ques. 38.

265. Emanuele (2018), "NDE 8722," Answer to Ques. 27.

266. Bridget F. (1995), "NDE 3648," Answer to Ques. 37, 39.

267. Sam Parnia, MD, PhD, and Tara Keshavarz Shirazi, BS, "What is the Best Available Evidence for the Survival of Human Consciousness After Permanent Bodily Death?" (Bigelow Institute for Consciousness Studies, contest entry 2021), 5-6.

268. Raymond A. Moody, MD, *Life After Life* (New York: Bantam 1975).

269. Kenneth Ring, PhD, *Life At Death: A Scientific Investigation of the Near-Death Experience* (New York: Coward, McCann & Geoghegan 1980).

270. Clinical death is defined by cardiologist and renowned NDE researcher Pim van Lommel, MD, as being unconscious, having stopped breathing, and with no palpable pulse or blood pressure. See his 2021 contest submission to the Bigelow Institute for Conscious Studies, titled: "THE CONTINUITY OF CONSCIOUSNESS: A concept based on scientific research on near-death experiences during cardiac arrest" (Bigelow Institute for Consciousness Studies, contest entry 2021), 7.

271. Bruce Greyson, MD, "The Near-Death Experience Scale: Construction, Reliability, and Validity," *The Journal of Mental and Nervous Disease,* Vol. 171, No. 6 (1983).

272. A June 2021 version of the Greyson Scale appears on the IANDS website at https://iands.org/research/important-research-articles/698greyson-nde-scale.html.

273. Parnia, "What is the Best Available Evidence," 7.

274. Jeffrey Long, MD with Paul Perry, *Evidence of the Afterlife: The Science of Near-Death Experiences* (New York: HarperCollins Publishers 2010). Quoted with permission of HarperCollins Publishers.

275. Jeffrey Long, MD with Paul Perry, *God and the Afterlife: the Groundbreaking New Evidence for God and Near-Death Experience* (New York: HarperCollins Publishers 2016).

276. Bruce Greyson, MD, *After: A Doctor Explores What Near-Death Experiences Reveal about Life and Beyond* (New York: St. Martin's Essentials 2021), 139.

277. Raymond A. Moody, MD, PhD. 2005. *The Light Beyond* Rev. Ed. (London: Rider Books 2005), xv. (Orig. pub. 1988.)

278. Greyson, *After,* 74-75.

279. Many out-of-body (OOB) souls describe their surroundings or state of existence as another dimension, world, universe, level, or even the afterlife itself. Knowings I received in the afterlife explained that OOB souls stay in the physical universe portion of Source's mind. They are simply temporarily untethered from their human body. However, because these souls have not, in fact, entered the afterlife and have not processed through the transition from human to Light Being perspective and thinking, OOB souls still think and believe as their human hosts do. The wealth of new information accessible to them is not Knowings, which can be received only in the afterlife, but is the soul's own database of everything it has recorded during that incarnation. We souls record every moment of life, including all sensory input, and also all childhood and adult thoughts, beliefs, religious training, and misunderstandings; all dreams; all fact and fiction to which the soul has been exposed; all of the soul's fantasies; and all jokes the soul has heard, among other categories of life experience. This information, according to the Knowings, cannot be relied upon as true because it is the soul's synthesis of human facts and fiction, not eternal truths downloaded from Source.

280. Ring 1982. *Life at Death,* 68.

281. Long, *Evidence of the Afterlife,* 3.

282. Greyson, *After,* 90, 78-89.

283. Karl L.R. Jansen, MD, "The Ketamine Model of the Near-Death Experience: A Central Role for the N-Methyl-D-Aspartate Receptor," *Journal of Near-Death Studies* 16 (1) (1997); Greyson, *After,* 110.

284. Greyson, *After,* 94.

285. Ibid., 95.

286. Ibid., 96-97.

287. Long, *God and the Afterlife,* 3-4. They did in fact meet the same type of being, what we call a Light Being. But they were all different Light Beings. Quoted with permission of HarperCollins Publishers.

288. Parnia, "What is the Best Available Evidence," 49-50.

289. Pim van Lommel, MD, "THE CONTINUITY OF CONSCIOUSNESS: A concept based on scientific research on near-death experiences during cardiac arrest" (Bigelow Institute for Consciousness Studies, contest entry 2021), 7.

290. Greyson, *After,* 152, 161.

291. Jeffrey Long, "Evidence for Survival of Consciousness in Near-Death Experiences: Decades of Science and New Insights" (Bigelow Institute for Consciousness Studies, contest entry 2021). A free PDF of the paper can be downloaded by clicking on Dr. Long's listing posted at https://www.bigelowinstitute.org/index.php/bics-afterlife-proof/bics-essay-contest-winners-runners-up/.

292. For an in-depth discussion of empathic or shared NDEs, see Raymond Moody, MD, *Glimpses of Eternity: Sharing a Loved One's Passage From This Life to the Next* (Paradise Valley, AZ: Sakkara Productions 2016).

293. Moody, *Life After Life,* 24.

294. Greyson, *After,* 48.

295. Moody. 2005. *The Light Beyond,* xiv.

296. Greyson, *After,* 11.

297. Nanci L. Danison, *BACKWARDS: Returning to Our Source for Answers* (Columbus, OH: AP Lee & Co. 2007); Nanci L. Danison, *BACKWARDS Guidebook* (Columbus, OH: AP Lee & Co. 2009); Nanci L. Danison, *BACKWARDS Beliefs: Revealing Eternal Truths Hidden in Religions* (Columbus, OH: AP Lee & Co. 2011); Nanci L. Danison, *Answers From The Afterlife* (Columbus, OH: AP Lee & Co. 2016); and Nanci L. Danison, *Create a New Reality – Move Beyond Law of Attraction Theory* (Columbus, OH: AP Lee & Co. 2018).